WORLD BANK STAFF WORKING PAPERS
Number 726

Macroeconomic Effects of Efficiency Pricing in the Public Sector in Egypt

Sadiq Ahmed
Amar Bhattacharya
Wafik Grais
Boris Pleskovic

The World Bank
Washington, D.C., U.S.A.

Copyright © 1985
The International Bank for Reconstruction
and Development / THE WORLD BANK
1818 H Street, N.W.
Washington, D.C. 20433, U.S.A.

All rights reserved
Manufactured in the United States of America
First printing April 1985

This is a working document published informally by the World Bank. To present the results of research with the least possible delay, the typescript has not been prepared in accordance with the procedures appropriate to formal printed texts, and the World Bank accepts no responsibility for errors. The publication is supplied at a token charge to defray part of the cost of manufacture and distribution.

The World Bank does not accept responsibility for the views expressed herein, which are those of the authors and should not be attributed to the World Bank or to its affiliated organizations. The findings, interpretations, and conclusions are the results of research supported by the Bank; they do not necessarily represent official policy of the Bank. The designations employed, the presentation of material, and any maps used in this document are solely for the convenience of the reader and do not imply the expression of any opinion whatsoever on the part of the World Bank or its affiliates concerning the legal status of any country, territory, city, area, or of its authorities, or concerning the delimitation of its boundaries, or national affiliation.

The full range of World Bank publications, both free and for sale, is described in the *Catalog of Publications;* the continuing research program is outlined in *Abstracts of Current Studies.* Both booklets are updated annually; the most recent edition of each is available without charge from the Publications Sales Unit, Department T, The World Bank, 1818 H Street, N.W., Washington, D.C. 20433, U.S.A., or from the European Office of the Bank, 66 avenue d'Iéna, 75116 Paris, France.

Sadiq Ahmed is an economist in the Egypt Division of the World Bank's Europe, Middle East, and North Africa (EMENA) Regional Office. Amar Bhattacharya is an economist in the office of the vice president, East Asia and Pacific Regional Office. Wafik Grais is an economist in the Yugoslavia Division of the EMENA Regional Office. Boris Pleskovic, an economist, is a consultant to the World Bank.

Library of Congress Cataloging in Publication Data

```
Macroeconomic effects of efficiency pricing in the public
   sector in Egypt.

   (World Bank staff working papers ; no. 726)
   Bibliography: p.
   1. Price regulation--Egypt.  2. Egypt--Economic policy.
I. Ahmed, Sadiq.  II. Series.
HB236.E3M33  1985         339'.0962        85-5384
ISBN 0-8213-0521-2
```

Abstract

At the mid-point of the 1980s, Egypt is characterized by intervention and regulation in all facets of economic activity with the government controlling some prices and directly determining some quantities. These policy-induced economic variables have led to distortions that affect the efficient allocation of resources and create misperceptions about the actual opportunity cost of goods and services. One result is a misallocation of resources and resultant heavy costs to the economy. During the period of the seventies, external resources were abundant, and the economy was able to sustain a high rate of economic growth despite the misallocation, whose costs were not perceived as a heavy burden. In the eighties, however, the world environment has been less favorable, and the inflows of external resources have been and are likely to be more limited. Hence concern over the cost of the misallocation of increasingly scarce resources is growing.

To analyze the relation between regulation, distortions and macroeconomic performance, measures of the levels of distortion are needed, a requirement that implies some evaluation of opportunity costs. This paper presents an economy-wide framework, called MISR2, that provides a general equilibrium evaluation of these costs and hence of distortions.a/

The MISR2 framework is designed to analyze the short- to medium-term macroeconomic consequences of policies aimed at minimizing the costs of the misallocation and distortions. The framework was used to analyze two sets of policy interventions: (i) adjustments of regulated prices and quantities within the same policy regime, that is, with the regulated variables remaining

---

a/ MISR is the Arabic name for Egypt. MISR2 is a more disaggregated and technically advanced follow-up to MISR1, which is a very simple aggregative model using the Transaction Value (TV) approach.

controlled; and (ii) adjustments in the policy regime itself, an approach that implies freeing controlled prices and quantities and letting market forces determine them. In MISR2 the focus is on four areas of economic intervention: (i) pricing of output in the public sector, (ii) employment and wage policy in the public sector, (iii) the trade and exchange rate regime; and (iv) rationing and subsidization as relates to households. In this paper, MISR2 is used mainly to analyze the effects of adjustments in public sector prices and in the regimes governing trade and exchange rates.

MISR2 disaggregates production into nine activities. Each production activity can be further separated into two, based on whether production takes place in the public or private sector. The distinction is needed on the one hand to capture the different pricing rules governing the output of the two sectors, and on the other hand to trace the relation between the pricing of public sector output and the mobilization of savings. Households are separated into two categories: urban and rural. This distinction reflects the different nature of the labor market of the two groupings. The MISR2 framework also captures the government as an independent agent that collects taxes, spends, saves and invests. The government also intervenes in various markets by supporting certain prices and subsidizing certain commodities. Finally, the rest of the world is separated into three groups in order to capture the three pools of foreign exchange through which transactions with abroad take place.

In order to obtain a quantitative evaluation of distortions, for each market MISR2 distinguishes between the actual price and the opportunity cost of the commodity concerned. The opportunity cost is determined for each commodity based on the functioning of the market and the nature of intervention. A measure of distortion in the market is obtained by calculating the difference between the values of transaction at the actual price and the opportunity cost respectively.

A predominant result of increases in prices in the public sector is a contraction of the economy. Raising prices is the same as imposing a tax on private sector income. When no other policy intervention takes place, the price increases generate an excess supply of resources in the public sector that lowers total absorption in the economy. In other words, raising public sector prices alone results in a trade-off between growth and efficiency. While distortions will be less, that result will occur at the expense of a slowdown in economic growth. The trade-off, however, can be improved by combining price increases with demand management policies.

Another important result is that a program of reform which liberalizes the economy with respect to the determination of prices in the markets for public sector output, to wage and employment decisions in the public sector and to determination of the exchange rate, will lead Egypt to a high-growth path while at the same time improving the macroeconomic balance considerably. A major reason is the improved efficiency of the public sector. In an environment riddled with major policy induced constraints, the actual productivity of key resources, foreign exchange, labor and capital, are much below their true levels. Consequently, when the distortions are removed, the realized productivity gains are substantial. The factors of production are now mobile and respond to market signals which now better reflect true economic values. Resources move from low productivity activities to high productivity activities enabling both higher output and lower costs. The macroeconomic balance also improves. Liberalization of the exchange rate combined with improved efficiency in the public sector augments the growth of exports, which in turn play a key role in sustaining high economic growth.

With improved earnings from exports, the current account deficit also narrows. The fiscal deficit falls as tax revenues increase because of the higher economic growth, and the fiscal contribution of the public sector picks up because of the price liberalization.

A further interesting result is that there is no necessary contradiction between price flexibility and a dampening of inflationary pressures. Price flexibility, by raising public savings, will contribute both to improved efficiency of resource use and reduced budget deficits because of greater public savings. Nominal aggregate demand will fall, whereas real aggregate supply will rise, causing a decline in inflationary pressures.

As expected, in a flexible environment distortions are virtually eliminated. Economic agents operate along their true supply/demand curves and resource allocations are governed by price signals which reflect true economic values. Consequently, the sources of rents no longer exist. In a flexible environment the dualism between public and private sectors does not exist since both face the same set of prices and operate under identical policy framework.

Acknowledgement

This paper is one of the final output of a research project (RPO 672-25A) organized jointly with the Bank, the USAID and the Development Research and Technical Planning Center (DRTPC) of the Cairo University. The authors would like to express their gratitude to all the members of the DRTPC group, but specifically to Amr Mohieldin, Mahmoud Fadil, Mohtaz Khorshid, Mohamed Osman and Ahmed Safti, for their valuable collaboration in developing the basic Social Acccounting Matrix (SAM). Many people contributed to the formulation of the analytical framework, MISR2. They include Graham Pyatt, Arne Drud, Sweder van Wijnbergen, John Whalley and Lance Taylor. The authors gratefully acknowldge their contribution. Special thanks are also due to Kemal Dervis and John Wall for their support and encouragement. Gustavo Trevino provided valuable research assistance. Editorial assistance was provided by Ms. Whitney Wattris and very capable typing assistance was provided by Lu Oropesa, Meaza Wegayehu and Roxane Malikyar. The authors gratefully acknowledge their support. The errors in the paper are the sole responsibility of the authors.

TABLE OF CONTENTS

Page No.

Chapter I   INTRODUCTION.. 1 - 6

Chapter II   THE MISR2 FRAMEWORK FOR POLICY ANALYSIS............... 7 - 8

     II. 1   An overview of MISR2............................. 8 - 14
     II. 2   Distortions and Their Measurement in MISR2....... 14 - 20
     II. 3   Macroeconomic Trends in the Reference Path....... 20 - 27
     II. 4   The Reference Path Results....................... 27 - 30

Chapter III   SHORT-RUN RESPONSES TO INCREASES IN PUBLIC SECTOR
               PRICES.. 31 - 32

     III. 1   Macroeconomic Equilibrium and the Financing of
           the Fiscal Gap.................................. 32 - 36
     III. 2   Increasing Public Sector Prices................. 37 - 54
     III. 3   Aggregate Demand Policies....................... 55 - 66
     III. 4   The Effects of the Combined Package............. 67 - 72

Chapter IV   MEDIUM-TERM PERSPECTIVE OF A PUBLIC SECTOR
               REFORM PROGRAM.. 73 - 74

     Phase I

     IV. 1   Medium-Term Response to Controlled Adjustments... 74 - 80

     Phase II

     IV. 2   Macroeconomic Trends in a Flexible Policy
          Environment...................................... 81 - 87

Chapter V   SUMMARY... 88 - 89

ANNEXES .... 90 - 199

REFERENCES .... 200 - 201

Chapter I

INTRODUCTION

Egypt's development effort has reached a difficult stage of transition. While growth in the 1970s was very rapid growth, in 1981/82 the expansion of output, consumption and investment slowed significantly (see Table 1.1). The medium- to long-term outlook is that, unless substantial policy measures are implemented without delay on numerous fronts, the slowdown will culminate in a long-term decline. The growth rate could drop to only

Table 1.1: Growth Trends, 1974-82/83
(percent a year)

	1974-80/81	1980/81-82/83
GDP at factor cost	9.1	7.3
Consumption	6.8	6.5
Public	6.8	10.0
Private	6.9	5.0
Investment	13.0	5.0
Public	18.2	4.5
Private	11.2	6.0

Source: Data provided by Egyptian authorities and World Bank staff reports.

about 4 percent a year in the 1990s. Such low growth is neither desirable nor even feasible for Egypt's development, since it will lead to growing unemployment and social problems. Egypt must continue to maintain strong investment growth and to improve the efficiency of investment to expand and renew its infrastructure and provide food, housing and employment to a population that continues to grow rapidly, at 2.8 percent a year.

The high economic growth in the 1970s was chiefly the result of a surge in foreign exchange resources, based mainly on exogenous factors--concessional capital inflows, primarily from other Arab countries, earnings from petroleum exports and from the Suez Canal, tourism receipts and workers' remittances (see Figure 1). While these resources enabled Egypt to achieve a high level of growth, they also made the economy very susceptible to outside influences. Further, by enabling Egypt to enjoy rapid growth without much stress on the domestic economy, they also helped policy-makers avoid difficult choices concerning the trade-offs between consumption and investment, the need to mobilize domestic resources to finance development and the need to improve the foreign exchange capability of the economy. Moreover, the growth was lopsided: sectors having very weak linkages with the rest of the economy, e.g., petroleum, and the Suez Canal, grew very fast, while domestic-based production sectors such as agriculture and industry grew relatively slowly (see Table 1.2).

During the same period, government expenditures also grew very rapidly. To a large extent they were financed through bank borrowing, with a consequent rapid expansion in the money supply. The high growth in income and money supply caused a rapid increase in nominal domestic demand. The resultant pressures from the excess demand were mitigated through imports and price increases. However, in order to moderate the impact of the resulting inflation of the lower income groups, the government maintained a strict policy of price controls on a wide range of goods produced by the public sector.

Figure 1.1: Development of Exogenous Resources

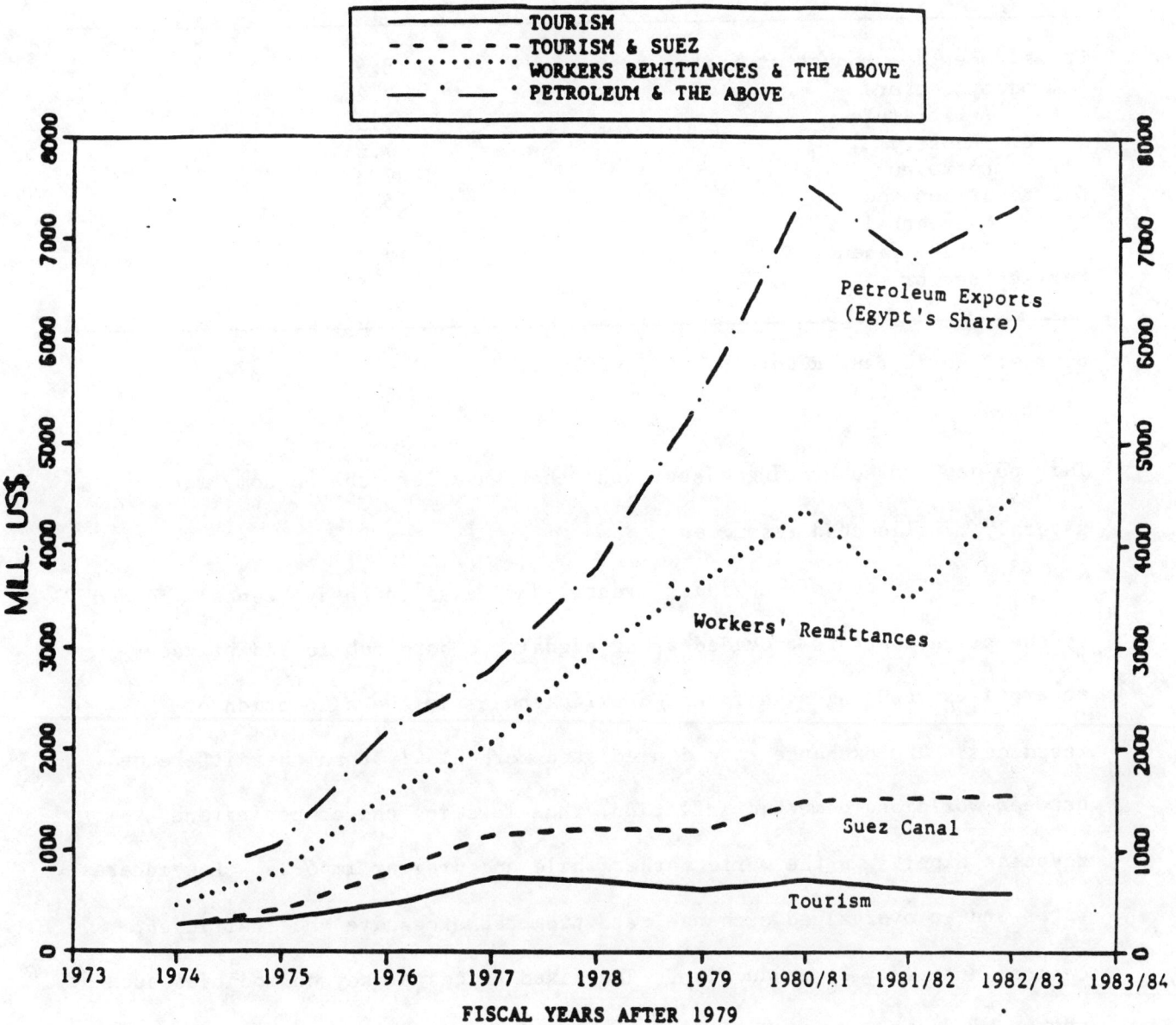

Table 1.2: Growth in Gross Domestic Product, 1974-1981/82
(annual percent)

Gross domestic product at factor cost	8.9
Commodity sectors	7.2
Agriculture	3.0
Industry	7.6
Petroleum	27.3
Distributrion sectors	15.3
Suez canal	214.7
Trade, finance and insurance	12.2
Service sector	6.5

Source: World Bank Report No. 4498-EGT, 1983

This policy caused growing distortions that have left the economy with severely misallocated resources.

The system of subsidies (rents) and taxes (negative rents) induced by the price controls provided wrong signals to both public and private enterprises, causing significant inefficiencies in the allocation of resources. The exchange rate depreciated more slowly than the difference between world and domestic inflation, thus lowering the competitiveness of Egyptian exports in the world market while encouraging imports. Low interest rates and an overvalued exchange rate promoted excessive and inefficient capital intensity of production. The fixed prices of key commodities such as energy products, wheat and flour promoted inefficient consumption, leading to waste, loss of foreign exchange and high budget deficits. Since the burden of fixed prices fell largely on the commodity-producing sectors (industry and agriculture), resources shifted from those activities to trade and services, which were more profitable. Within the commodity-producing sectors, price controls and subsidies induced an inefficient pattern of production by encouraging activities in which Egypt lacked comparative advantage (aluminum, iron and steel, basic chemicals and meat production) at the expense of activities that should have been promoted more vigorously (cotton textiles, manufactured food, cotton, fruits and vegetables (World Bank, 1983).

Apart from inducing a misallocation of resources, the pricing policy also contributed to a weakening of the fiscal situation. The overall impact of the price controls on the financial performance of public enterprises was very negative (Ahmed, 1984). As a consequence, public enterprises frequently had to resort to the treasury for financial support, further augmenting the budget deficit and leading to an increase in the supply of money and in inflationary pressures. Thus, paradoxically, price controls helped sustain high inflation rather than reducing it.

The lesser flow of external resources since 1981 produced a slowdown in economic growth. The downturn was triggered by a fall in oil prices, exacerbated by a decline in tourism receipts and remittance earnings and a slowdown in Suez Canal earnings, following the general slowdown in the world economy (see Figure 1.1). The prospects for greater external resources are not too bright. Therefore, the challenge that Egypt now faces is to make those difficult choices it was able to avoid when exogenous resources were plentiful. They involve: the mobilization of resources, raising the efficiency of investments, improving the capability of the domestic economy to absorb a growing labor force; and earning adequate foreign exchange. Economic growth will depend on the pace and comprehensiveness of the policy reforms which should be designed to eliminate the distortions in the economy. The outcome will be improved efficiency and productivity of investments along with greater mobilization of domestic resources to finance Egypt's investment needs and to help restructure the economy so as to reduce the dependence on exogenous sources of growth and foreign exchange earnings.

Numerous policies have combined to sustain the pervasive distortions in the Egyptian economy. Of them, price controls (including control over the exchange rate) appear to be the most distortive. They are largely responsible for the inefficiency of investments in both industry and agriculture (World Bank, 1983). Moreover, as mentioned above, pricing policy is an important cause of the large fiscal deficits Egypt has experienced: the direct and indirect subsidy imparted by controlled prices exceeds one-fifth of current GDP. Reform of the pricing policy should, therefore, constitute a major component of any remedial program. Price liberalization alone, however, will not necessarily improve all indicators. The impact of other distortions on certain indicators might even be strengthened if only public sector prices are adjusted and the other distortions are left intact. What is needed, therefore, is a comprehensive policy package that includes price liberalization as an active element.

The main objective of this paper is to study the macroeconomic effects of price liberalization in Egypt. Chapter 2 contains an overview of the macroeconomic framework used in the analysis, MISR2, a description of the main distortions it captures, the measurement of these distortions and a macroeconomic outlook for Egypt for the 1980s under the policy regime prevalent in mid-1983. Chapter 3 presents an analysis of the short-run (within a year) macroeconomic response of the Egyptian economy to a package of policy measures that includes increases in public sector prices. In Chapter 4 a medium-term--10-year--public sector reform program is described. During the first five years public sector prices, although still controlled, are raised in order to reflect opportunity costs better. During the later five years public sector prices are freed and other distortions removed. Chapter 5 offers a summary of the main findings. Detailed annexes describe the technical aspects of the macroeconomic framework.

Chapter II

THE MISR2 FRAMEWORK FOR POLICY ANALYSIS

As stated in Chapter I, Egypt is now characterized by intervention and regulation throughout the economy, with the government controlling prices and setting some quantity targets. Price and quantity controls create distortions in the sense that goods and services are not valued at their opportunity cost. The distortions in turn affect the efficient allocation of resources and have macroeconomic consequences on economic activity. The outcome has been heavy costs to the economy. During the 1970s, these costs were not perceived as a major economic burden: external resources were abundant, and the economy was able to sustain a high rate of economic growth. The world environment in the eighties, by contrast, has been less favorable with inflows of external resources likely to be more limited. Hence there is a growing concern over the cost of the misallocation of increasingly scarce resources.

In order to analyze the relation between regulation, distortions and macroeconomic performance, measures of the levels of distortion are needed. As such, the opportunity costs must be evaluated. The MISR2 framework provides a general equilibrium evaluation of these costs and distortions. It also is designed to analyze the short- to medium-run macroeconomic effects of policies aimed at lessening the costs of misallocation and distortion. The framework is set up to analyze two sets of policy interventions: (1) adjustments of regulated prices and quantities within the same policy regime, that is, the economic controls remaining as they are now; (2) adjustments in the policy regime itself that would free controlled prices and quantities and allow market forces to determine them. The focus in MISR2 is on four areas of economic intervention: (1) the pricing of output in the public sector; (2) public sector employment and wage policy; (3) the trade and exchange rate regimes; and (4) rationing and subsidization of households. In this paper, MISR2 is used to look at the effects of adjustments in public sector prices and in the trade and exchange rate regimes.

This chapter provides an overview of the MISR2 framework and the main distortions it captures. Those distortions are then defined and measured. Finally, MISR2 is used to derive a macroeconomic outlook for Egypt for the 1980s under the policy regime prevalent in mid-1983, along with the accompanying levels of distortions. This outlook is not forecast. Rather, it represents a view of the economy based on assumptions about the world environment, policy interventions and the behaviors and institutional arrangements governing the Egyptian economy.

II.1 AN OVERVIEW OF MISR2

MISR2 is a multisectoral, two-household, economy-wide framework of the Egyptian economy.[1] It disaggregates production into nine activities, with each production activity separated into two where appropriate, depending on whether production takes place in the public or private sector. This

---

[1] a full account of the economics and specifications of MISR2 is given in the appendices.

distinction is needed on the one hand to capture the different pricing rules governing output in the two sectors and on the other hand to trace the relation between the pricing of public sector output and the mobilization of savings. Households are separated into two categories--urban and rural--reflecting the different nature of their labor markets. The MISR2 framework also captures the government as an independent agent that collects taxes, spends, saves and invests. The government also intervenes in various markets by supporting certain prices and subsidizing certain commodities. Finally, the rest of the world is separated into three groups to account for the three pools of foreign exchange through which transactions with abroad take place.

MISR2 is a sequential equilibrium economy-wide model. It allows a user to derive a path for the economy as a sequence of equilibria. The equilibrium defined in each period is the outcome of the behavior of economic agents and institutional arrangements in the economy. The latter ensure the overall consistency of the independent decisions of the agents. The equilibrium for each period within the 10-year timeframe is also dependent on a set of conditions related to agents, policy intervenions, the world envrionment and the endowments of the agents. These conditions are allowed to change between periods according to behavioral and accounting rules or exogenous trends.

In MISR2 the price of the output from each public sector activity is regulated, and pricing is considered a policy instrument in the hands of the government. One question is how the markets for these commodities clear. For all markets except those for electricity, oil and services, producers are assumed to maximize profits at the regulated price. Thus, they define the level of supply associated with that price. The demand for exports is satisfied first, taking its share of the supply. The remaining supply goes to the domestic market. Rationed in this way, demand has to adjust and does so by an upward shift in the regulated price. That change will first produce: (1) an expansion of supply; (2) a dampening of the demand for exports; (3) a lower rationing of domestic demand. These responses of course depend on the elasticities of supply and export demand. In the electricity and service sectors, producers are assumed to satisfy whatever demand exists at the regulated price. Their supply is thus perfectly elastic at that price. There is a wedge between the regulated market price and the marginal cost. If demand is beyond the supply that results in maximum profit, producers are "taxed". Otherwise they receive a pure profit above the rent associated with the maximum profit. An upward adjustment in the price of electricity or services affects production through variations in demand and has direct implications on the profitability of the sector. The oil market behaves differently. The domestic price and output are fixed, and export demand is perfectly elastic, conditions that allow the market to clear. An upward adjustment in the price of oil tends to contract dometic demand and to leave a larger exportable surplus.

Public sector wages, whether in urban or rural areas, are fixed and considered as policy variables. The rural wage in the private sector is assumed to clear the rural labor market whereas the private wage in the urban sector responds between periods to unemployment and expectations of inflation. Thus an endogenous wage differential between public and private sector wages is allowed for. Further, it is assumed that migration from rural to urban areas takes place in between periods according to the expected urban-rural wage differential.

The MISR2 framework captures another feature of the labor markets in Egypt. The government expects public sector activities to absorb a certain amount of the increase in the labor force. Thus, they do not have much leeway in their hiring decisions. Furthermore, they cannot fire labor except under exceptional circumstances. Thus, for all practical purposes their wage bill is fixed. This rigidity in public sector employment is reflected in MISR2 by assuming that labor in public sector acitivities is a fixed factor. In this case a rent that may fall short or exceed the wage bill accrues to labor. The difference between the rent and the cash payment is absorbed through the operating surplus of the respective acitivities. In practice, this amounts to a tax imposed on production in the public sector.

The MISR2 framework captures the trade regime in Egypt by assuming three balance of payments with three exchange rates. The first balance of payments--the central bank pool of foreign exchange--has a fixed exchange rate and clears via net transfers of foreign exchange to the second balance of payments--the commercial banks' foreign exchange pool. Most imports that go through the central bank pool are restricted by quotas. The first balance of payments thus works like a traditional balance of payments with a fixed exchange rate and adjustments in net borrowing. The only difference is that the net borrowing from abroad is assumed to be set by policy, and the adjustment happens through net borrowing from the commercial banks' foreign exchange pool. This latter pool also has a fixed exchange rate and fixed net borrowing from abroad. Here, the supply and demand for foreign exchange clear via a rationing of imports. Once compulsory transfers are made, there is a certain amount of foreign exchange left, which is then allocated across various commodities. This allocation, jointly with the world prices and the exchange rate, determines the amounts of imports of each commodity through the second poool. The third balance of payments corresponds to the parallel market of foreign exchange. Here the exchange rate adjusts to equilibrate the inflows and outflows of foreign exchange.

There are three types of interactions between the three pools. The first one, already mentioned, is the net borrowing or lending between the central bank and commercial bank pools. A tightening of the foreign exchange resources in the first pool leads to increased rationing of the imports coming through the second pool, as the amount of foreign exchange available there is reduced. Another interaction occurs through a spillover of imports from the first and second pools into the third pool that drives up the parallel market exchange rate. Thus excess demand for imports ultimately shows up a depreciation of the parallel market exchange rate. 2/ The third type of interaction is between the commercial banks and parallel pools: while the total amount of workers' remittances is given in dollars, their allocation between the two pools depends on the relative exchange rates. Greater depreciation of the parallel market exchange rate relative to the commercial bank rate results in a larger share of remittances going to the parallel market.

Equity considerations have led Egyptian policy-makers, since the early sixties, to establish a system of distribution of basic commodities at controlled prices. More or less the same pertains in the early eighties. It is captured in MISR2 by providing agriculture, food processing and other industry commodities at subsidized prices to various users. The commodities are not, however, available in unlimited quantities, and users are assumed to face some rationing. In particular, households are off their notional demand curves for these commodities. Increased availability of the commodities at the controlled prices means an increased level of subsidization and a deterioration in the budget deficit.

2/ However, because of institutional rigidities, imports are not allowed to shift completely to the third pool without costs (see appendices).

One essential feature of the Egyptian economy is the duality between the public and private sectors. That duality is reflected in the macroeconomic closure of MISR2. There are two pools of savings, one for the private, the other for the public, sector. All the savings of the agents of the private sector are channeled to the private pool, which also receives foreign savings through the parallel market of foreign exchange. The latter savings are fixed in dollars and variable in domestic currency. The total resources of the private savings pool are channeled partly to private investors and partly to the public savings pool. This allocation is a fixed share that is typically a policy tool of the government. When a policy adjustment leading to an improved fiscal gap is implemented, there is a case for reducing the share of private funds channeled to th public sector. Indeed, a policy to improve public resource mobilization implies a transfer from the private to the public economy. If in addition, the public sector continues to borrow the same share of private funds, the resources of the private economy will be squeezed and a reduction in private activity will ensue. In other words, private investment is financially constrained. Thus, the financing policies the government follows to fill the public sector gap can crowd the private sector out.

The resources available to the public savings pool are composed of funds obtained from the private sector, the savings of the public sector and foreign savings coming through both the central bank and commercial bank foreign exchange pools. These foreign savings are fixed in US dollars and in domestic currency as the respective exchange rates are fixed. The resources of the public savings pool are used to finance public investment. The public sector investment budget is, however, a policy variable and is fixed nominally. Hence it may exceed or fall short of the resources available to finance it. In the absence of changes in the policy of the public sector relative to borrowing from domestic or foreign sources, an excess in public sector investment results in an expansion of activity that tends to pull in imports and to depreciate the parallel market exchange rate. Simultaneously, agents whose spendings are import-intensive without much scope for substitution face a deterioration in their real expenditures. This is particularly the case for the current and capital expenditures of the public sector. Thus excess public sector investment leads through the parallel market exchange rate to a shift from imported to domestic goods. At the same time it reduces the real value of import-intensive expenditures, lowering the components of absorption.

II.2 DISTORTIONS AND THEIR MEASUREMENT IN MISR2

In a market economy, prices are signals that allow agents to make economic decisions. Wrong signals are likely to lead to wrong decisions and a consequent misallocation of resources and overall economic loss.

What is the characteristic of a wrong signal and the resultant inappropriate price? Generally, an inappropriate price does not reflect the opportunity cost of the commodity it values. 3/ That definition raises two questions: what is the opportunity cost of a commodity and of how is it measured? 4/ The MISR2 framework deals with these questions as described below.

Consider the markets for commodities produced by the public sector. In all activities apart from electricity, services and oil, producers maximize their profits at the regulated prices, and domestic users are

---

3/ For a comparative review of price distortions and their relation to growth, see Agarwala (1983).

4/ Other questions are: what are the causes of the distortions and in which markets they appear?

rationed. Figure 2.1A shows the level of q_T of production and supply at the regulated price p. In Figure 2.1B part of q_T goes to exports, 5/ leaving q_d for domestic demand. The quantity q_d is the level of the ration. Given the domestic demand schedule D for public sector goods, the virtual price p_y (see Neary and Roberts, 1980) is the price that would clear the market at the level of the ration. In other words, users would be willing to pay p_y to acquire the level of ration q_d. But actually they pay the regulated price p. If the schedule D is derived from utility maximization, then the price p_v is a measure of marginal utility that is to the users of q_d a measure of its opportunity cost. Clearly from Figure 2.1B, the regulated price is different from the opportunity cost. The difference between the value of q_d at its opportunity cost and its regulated price is the shaded area, which is a measure of the distortion in the market. The MISR2 framework allows p_v and the shaded area to be computed. The latter is considered to be a rent accruing to the users. It goes back to their income and is reallocated with that income over all their outlays.

The markets for electricity and services work differently, as shown in Figure 2.2. In Figure 2.2A, the demand schedule is D, the marginal cost curve is at MC and firms are obliged to supply along the line S at the regulted price p. The market equilibrium quantity is at q_e. However, the marginal cost of producing q_e is at p_v; hence for that level of output producers are not getting their marginal cost, which is a measure of their opportunity cost for producing q_e. The regulated price p is different from the opportunity cost p_v, and the level of distortion in the market can be measured by the value of the shaded area. The MISR2 framework

---

5/ Besides exports, some purchases are made by the government trading authority, but these are not price responsive.

Figure 2.1

Supply Driven Markets of Public Sector Goods

2.1A. Production and Supply   2.1B. Domestic Demand

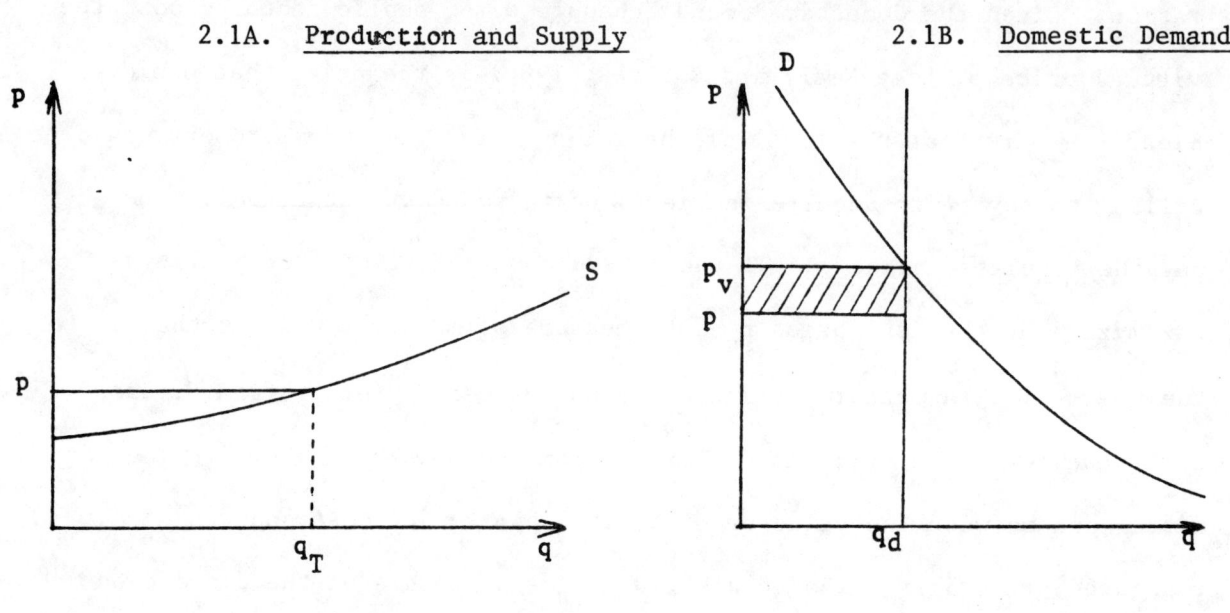

Figure 2.2

Demand Driven Markets of Public Sector Goods

2.2A. Demand Is Beyond Maximum Profit Supply   2.2B Demand Is Below Maximum Profit Supply

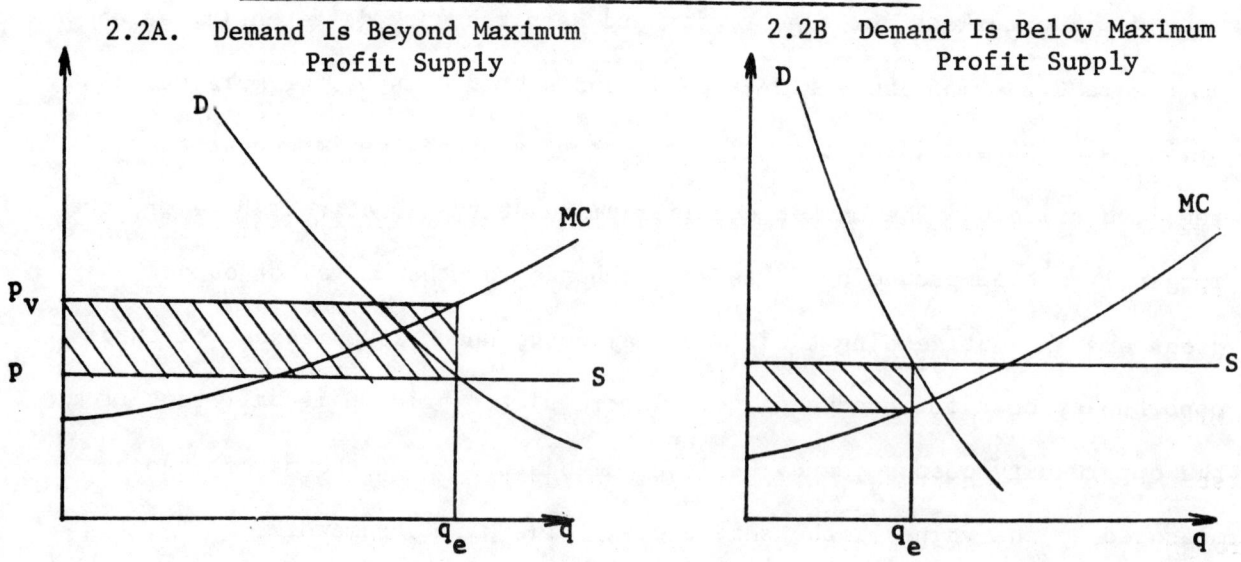

allows the marginal cost p_v and the difference in the value of q_e measured at its opportunity cost and at the regulated price to be computed. The shaded area is considered to be a negative rent to producers and is taken out of their operating surplus. Consequently, actual rate of return on their fixed assets is lower than if their output was measured at its opportunity cost. Figure 2.2B represents a situation where the regulated price is set too high, higher than the opportunity cost. Production is protected and is getting a pure profit, measured by the shaded area, but consumers are hurt.

Public sector employment is another area of government intervention that may lead to distortions. Because public sector firms have very little leeway in deciding their level of employment, labor is considered in MISR2 to be a fixed factor. At the same time, the wage the firms pay is a policy variable fixed by the government. Public sector firms are thus in a situation such as the one described in Figure 2.3. D is the schedule of the marginal productivity of labor, $\bar{L}$ is the labor force, and w represents the regulated wage. The marginal productivity of $\bar{L}$ that can be considered a measure of its opportunity cost is at w_v; firms pay a wage bill in excess of the opportunity cost value of the labor services they get equal to the shaded area. That bill is an additional negative rent on the firms that further cuts into their operating surplus and profitability. The shaded area in Figure 2.3 can be used as a measure of the distortion in the labor market faced by public sector firms.

All the agents intervening in the economy have demands for imports. In MISR2, imports come through the three foreign exchange pools. But in the first two pools - central bank and commercial bank - imports are rationed through quotas and the allocation of foreign exchange, respectively. The assumption is that agents have a demand for a composite of imports and that they allocate that demand through the three pools by minimizing the cost of obtaining the composite of imports. They thus have explicit demands for imports from each pool. 6/ Through the first two pools, the supply and demand for imports are as shown in Figure 2.4. The demand schedule is at D and imports are rationed at q_m. The users' oportunity cost for q_m is the marginal utility of q_m given by p_v, the virtual price. However, given the actual world price and the exchange rate, users are really paying p_m. Thus, the opportunity costs of imports through the two first pools of foreign exhange are different from the market prices charged for them. This distortion is measured by the shaded area shown in Figure 2.4. The value of the shaded area is a rent that accrues to the users of the imports. It goes into their income account and is allocated, together with the rest of the users' income, on various outlays.

One last area of government regulation captured by MISR2 is the provision of subsidized commodities to households, in particular agricultural goods, processed food and other industrial goods. The availability of these goods at subsidized prices is limited, and the markets faced by the households are like the one described in Figure 2.1B,. The ration is q_d and the subsidized price p. The price p_v measures the opportunity cost to households of the level of ration q_d. Again the shaded area can be used as a measure of distortion. It is a rent that accrues to households. This is captured in MISR2 by adding that rent to households' disposable income, which is reallocated over consumption and savings.

6/ This specification is adopted in order to reflect institutional constraints that do not allow agents to shift fully between the pools to meet their demand.

Figure 2.3

Regulation of Public Sector Employment

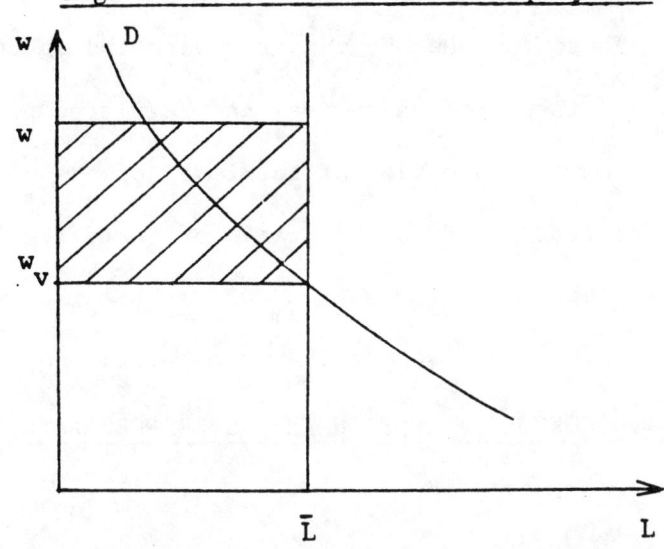

Figure 2.4

Imports through the Central Bank and Commercial Pools of Foreign Exchange

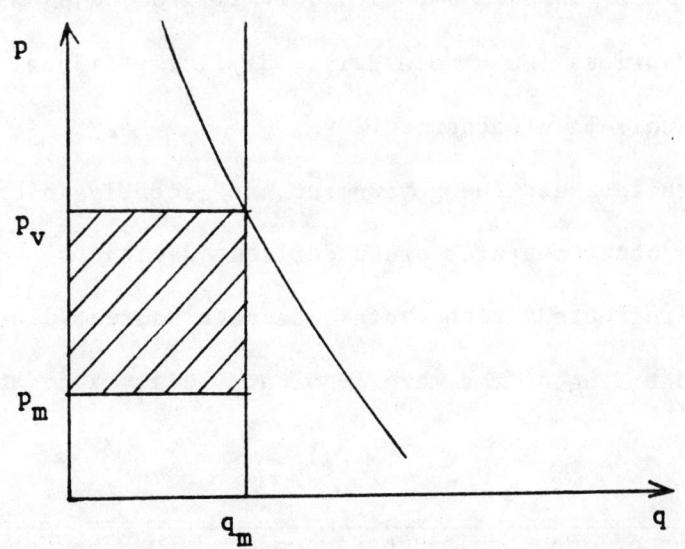

Thus in all markets where there is regulation, the MISR2 framework allows computation of the opportunity costs that result from the defined general equilibrium.7/ These can then be compared with the actual regulated prices. The difference in the opportunity cost and regulated price values of the goods traded is used as an indication of the level of distortion. In practice the share of the rent in GDP or the ratio of the virtual to the regulated prices can be used.

II.3 MACROECONOMIC TRENDS IN THE REFERENCE PARTH

This section presents the macroeconomic results of the reference path, which provides a benchmark for comparing the impact of policy changes on the various macroeconomic variables. In deriving the results of the reference path, it is assumed that policy regimes and interventions remain the same as in the past. Thus, the existing policy framework, with its price and exchange rate regimes, is assumed, with only limited adjustments in all government-controlled prices (see Table 2.1). Similarly, fiscal and monetary policies are assumed to remain unchanged.

In some of these areas the government has recently initiated certain policy reforms, while other measures are under consideration. None of these actions is, however, reflected in the reference path described below, mainly because the scope and timing of the more important policy initiatives remain uncertain.

7/ The virtual prices or opportunity costs computed in MISR2 are conditional on other distortions in the economy. Thus they could still be far from a "pure" opportunity cost where no distorition exists.

Table 2.1 - __Assumptions Underlying the Reference Path__

	1983/84	1984/85	1985/86	1986/87	1987/88	1988/89	1989/90	1990/1991	1991/92-1992/93
__Fiscal policy__									
Share of private savings mobilized by public sector	0.470	0.460	0.450	0.450	0.450	0.450	0.450	0.450	0.450
Nominal investment in state enterprises	0.230	0.250	0.230	0.210	0.200	0.180	0.180	0.150	0.140
__Pricing policy__									
Public food processing	0.075	0.075	0.060	0.060	0.060	0.060	0.060	0.060	0.060
Domestic price of petroleum products	0.100	0.110	0.120	0.120	0.120	0.120	0.120	0.120	0.120
Public textiles	0.090	0.075	0.060	0.060	0.060	0.060	0.060	0.060	0.060
Public other industries	0.090	0.075	0.060	0.060	0.060	0.060	0.060	0.060	0.060
Public electricity	0.050	0.050	0.050	0.050	0.050	0.050	0.050	0.050	0.050
Public transport and communication	0.100	0.070	0.090	0.085	0.085	0.085	0.085	0.085	0.085
Public services	0.100	0.090	0.080	0.080	0.080	0.080	0.080	0.085	0.080
Public construction	0.080	0.070	0.070	0.070	0.070	0.070	0.070	0.070	0.070
__Trade Policy__									
Commercial bank pool exchange rate	0.050	0.045	0.040	0.040	0.040	0.040	0.040	0.040	0.040
Tariff	0.0	0.0	0.0	0.0	0.0	0.0	0.0	0.0	0.0

The reference scenario is not a projection of Egypt's future. It simply reflects how the principal macroeconomic aspects of the ecnomy would evolve given the underlying assumptions.

3.a External Environment

As pointed out in Chapter 1, the major challenge facing policy-makers in Egypt emanates from the likelihood that the external factors which underlay the rapid economic growth since 1974 will be less favorable in th future. In defining the prospective external environment in the medium- and long-term, four factors are particularly important: petroleum and gas, workers' remittances, receipts from the Suez Canal and from tourism, and inflows of external capital. The key assumptions relating to these and other factors in the external environment are summarized in Table 2.2.

Table 2.2 - External Environment Underlying the Reference Path

	1983/84	1984/85	1985/86	1986/87	1987/88	1988/89	1989/90	1990/91-1991/92
Export price of oil (nominal growth rate $)	0.03	0.05	0.08	0.09	0.09	0.09	0.09	0.09
Oil production ($ growth rate)	0.06	0.07	0.06	0.055	0.05	0.04	0.035	0.03
Workers' remittances ($ growth rate)	0.13	0.095	0.09	0.085	0.085	0.085	0.08	0.08
Suez earnings ($ growth rate)	0.14	0.13	0.13	0.12	0.115	0.115	0.11	0.11
Other non-factor services ($ growth rate)	0.115	0.13	0.115	0.115	0.115	0.115	0.11	0.11
Foreign borrowings ($ growth rate)	0.17	0.17	0.16	0.15	0.13	0.11	0.10	0.10

Prospects for Oil and Gas

During the last several years Egypt has been successful in attracting large volumes of foreign investment in the oil sector, which has sustained a vigorous oil exploration and development effort. Although the success rate of this effort has been high, no new large discoveries have been made. Thus, the growth of oil production in the medium-term will depend on the pace of development of the newly discovered smaller fields and additional investments in secondary recovery to offset the declining production from the older, larger fields. In the longer term--beyond the mid-1980s--there is greater uncertainty about oil production because the projections of output have to be based on yet undiscovered fields. The incremental cost of finding and producing Egypt's oil will increase steadily, a trend that will have an adverse effect both on Egypt's share of the oil produced and on the attractiveness to oil companies of additional investment. On the whole, therefore, the outlook is that oil production will decline gradually, slowing considerably in the later years of 1980s.

Another issue is the international price of oil. There is a great deal of uncertainty about future price trends, given recent upheavals in the oil market. Current expectations are that oil prices will rise very slowly over the next two years and then recover somewhat in the medium-term. They are, however, unlikely to show large real increases.

The prospects for increasing natural gas output are very favorable, since reserves are relatively abundant and both associated gas and existing gas fields are presently underexploited. However, the substitution of gas for oil in energy use will require synchronized investments in the development of fields and related distribution networks, and in creating a capacity in the power and industrial sectors to use natural gas. In the medium term these investments are likely to constrain the potential utilization of natural gas.

Workers' Remittances

Workers' remittances have been another major source of external earnings for Egypt. With Egypt's increasing financial openness, remittances are now more susceptible to the political climate and financial conditions, as demonstrated by the large dip in remittances in 1981/82. Thus, in the short term remittances should show greater volatility. At the same time, the underlying trends that drive the medium term have also become less favorable. First, the rapid outmigration that characterized the 1970s appears to have slowed dramatically. With increased competition from other labor sources and slower income growth projected in the labor-importing countries because of reduced earnings from oil exports, the prospects for continued net outmigration are limited. Real income of migrants will, of course, continue to increase, but so will their standards of consumption. A great deal will also depend on political conditions and investment opportunities in Egypt compared with abroad. The most likely scenario is one of continued but slower real growth of remittances in the medium term.

Suez Canal Earnings and Tourism Receipts

Large increases in Suez Canal earnings also appear unlikely. Although tariffs have been raised substantially in recent years, further large increases will be difficult to achieve without eroding competitiveness. On the other hand, there is still potential for increasing the flow of traffic and raising the proportion of large vessels on the basis of the recent expansion. Nevertheless, even assuming a sustained expansion of the world economy, the pace of revenue growth will not match that of the 1970s.

Like the Suez Canal, the prospects for continued increases in <u>tourism receipts</u>, which are a major determinant of other exports of non-factor services, are only moderately favorable and are contingent on economic recovery in the industrial countries. Because of considerable investment, hotel capacity has greatly increased; however, other tourist facilities need to be upgraded. While tourism earnings are projected to increase steadily, once again they are unlikely to match past growth.

Flows of External Capital

As discussed in Chapter 1, the large increase in capital inflows during the second half of the 1970s allowed not only to finance a growing level of imports, but also to raise the level of foreign savings available for investment. A reversal of this trend has, however, been discernible since 1979, with a slowdown in official aid commitments and a rise in debt service payments.

Foreign investment within the Law 43 framework has slowed somewhat in recent years. In response, the government has introduced several measures to streamline investment. Assuming further efforts in this direction, Egypt should be able to attract large volumes of investment under the Law 43 framework. The present outlook, therefore, is one of continued modest growth in direct foreign investment.

Official aid flows will depend on the growth of new commitments and the disbursement rate. At present, roughly 35 percent of new commitments are in the form of commodity aid. The remainder consists largely of project loans, with virtually no direct balance of payments financing. Five sources account for almost 85 percent of official aid: involving four bilateral programs, Japan, the United States, the Federal Republic of Germany and France, and multilateral support from the World Bank. The growth in commitments from all these sources is expected to decelerate over the next few years and in some cases these commitments may decline in real terms. Therefore, it is unlikely there will be any large real increases in the growth in total aid commitments. One source of additional finances is renewed lending by the Arab countries. While this is possible, the levels are likely to be smaller than in the past. Despite the anticipated slowdown in the growth of commitments, a somewhat faster rate of increase of disbursements (i.e., actual inflows) should be possible based on more rapid utilization of foreign aid.

Nevertheless, net resource transfers from foreign aid will not show a significant increase because of the debt service requirements. Given the debt structure and rising level of outstanding debt, Egypt will have to finance increasing levels of both interest and amortization payments over the next five years.

Given Egypt's large import and development needs and the anticipated slowdown in foreign exchange earnings, net official aid and transfers, Egypt may have to rely more on non-concessional flows than in the past. This trend has already been discernible in the last two to three years. The shorter maturities and higher interest rates of these flows tend to limit the growth of commercial borrowing. Nevertheless, in the second half of the 1980s--when the slowdown in oil earnings is expected to more pronounced--Egypt may have to seek such additional financing to meet its balance of payments needs.

II.4 THE REFERENCE PATH RESULTS

Table 2.3 summarizes the growth scenario underlying the reference path. The average annual real rate of growth of GDP is only 5.9 percent in the reference path, compared with 9.4 percent during 1974-1981/82. This considerable slowdown is the result of both the poor investment performance and the growing level of distortions. Real investment rises by only 3.4 percent annually, compared with 12.6 percent during 1974-1981/82. The decline in investment growth reflects the difficulties associated with mobilizing resources. The slowdown in investment and income also has adverse effects on consumption and employment growth.

The difficulties encountered in mobilizing savings for financing investment are illustrated in Tables 2.4 and 2.5. The main reason for the decline in investment growth is the lower availability of foreign savings (Table 2.5). Domestic savings fail to offset the impact of this slowdown. Thus public savings remain flat, while private savings show only marginal improvement. The disaggregation of public savings indicates government savings continue to be negative, while public company savings are virtually unchanged, despite the significant premium obtained from the exchange rate system. The overall fiscal deficit remains very high (Table 2.6). Budgetary difficulties are linked to the decline in foreign resources, since a major

Table 2.4 — **Growth Scenario in the Reference Path**
(real rates of growth per year)

	Historical average 1974-1981/82	Reference path 1982/83-1991/92
Household consumption	7.5	5.2
Government consumption	8.7	3.4
Total consumption	7.9	4.8
Public sector investment	10.9	3.3
Private sector investment	17.4	3.6
Total investment	12.6	3.4
Exports (gnfs)	12.7	5.9
Imports (gnfs)	7.2	3.0
GDP (at mp)	9.4	5.3
Employment	3.3	2.3

Table 2.5 — **Savings-Investment Balance in the Reference Path**
(as a share of GDP)

	Historical		Reference Path	
	1980/81	1982/83	1987/88	1991/92
Total investment	32.1	27.5	28.1	28.5
Household savings	-	12.1	12.8	12.6
Private company savings	-	3.3	3.6	3.5
Total private savings	14.3	15.4	16.4	16.1
Public company savings	-	6.7	6.6	6.8
Government savings	-	(-)5.7	(-)4.5	(-)3.6
Social security savings	-	4.7	4.7	4.7
Total public savings	5.2	5.7	6.8	7.9
Total foreign savings	12.6	5.8	4.9	4.5

Table 2.6 - Selected Fiscal Indicators in the Reference Path
(as a share of GDP at market prices)

	Actual 1982/83	Reference Path 1987/88	Reference Path 1991/92
Total current expenditure	21.3	19.0	18.6
Subsidies	7.6	6.0	5.3
Total tax receipts	21.7	19.4	19.1
Public economic sector surplus	8.8	7.0	5.6
Total public investment	18.8	19.2	19.4
Overall fiscal resource gap	17.3	16.8	16.2
Foreign borrowings	5.5	4.6	4.1
Social security savings	4.7	4.7	4.7
Domestic private financing	7.1	7.5	7.4

component of budgetary revenue depends significantly on petroleum and Suez Canal earnings, which are now much less favorable. The small decline in the overall fiscal deficit is predicated on the maintenance of the present program of expenditure control. Based on this assumption, both current expenditures and subsidies fall as a percentage of GDP.

Export performance declines sharply from 12.7 percent during 1974-1981/82 to 5.9 percent in the reference path. This is primarily attributable to a fall in oil exports. Non-oil exports do not show much improvement, partly because of incentives, but chiefly because of the inefficiency of production. The depreciation of the parallel market exchange rate is much less than the differential between domestic and world inflation, causing a deterioration in the international competitiveness of Egyptian exports (see Table 2.7).

More importantly, a major problem highlighted in the reference path is the growing impact of distortion on the Egyptian economy (Table 2.8). The misallocation of resources induced by these distortions is an important factor contributing to the decline in Egypt's growth prospects in the reference path.

2.7 - Key Price Developments in the Reference Path
(percentage annual change)

	Reference path (1982/83-1991/92
Wages	
Public sector (urban)	12.1
Private sector (urban)	13.9
Public sector (rural)	12.1
Private sector (rural)	17.0
Prices	
Consumer (urban)	12.2
Consumer (rural)	13.4
Investment	15.3
Exchange rate	
Commercial bank pool	4.2
Parallel market	8.7
Memo item: International inflation	6.5

Table 2.8 - Indicators of Distortions in the Reference Path
(as a share of GDP)

	1982/83 (Actual)	1987/88	1991/92
Share of rents in GDP			
Subsidized consumer goods			
Rents to urban households	0.49	0.69	0.67
Rents to rural households	1.21	1.77	1.78
Total rents	1.70	2.46	2.45
Public sector employment	1.10	1.96	1.90
Imports Premia	4.2	9.3	11.0
Public sector output market			
Supply-driven sectors	14.1	19.8	19.7
Demand-driven sectors	1.3	0.98	0.8

Chapter III

SHORT-RUN RESPONSES TO INCREASES IN PUBLIC SECTOR PRICES

One purpose in adjusting public sector prices is to improve the incentives in the public sector and allow for better resource mobilization. This expectation and the likelihood of lower distortions and higher efficiency in the economy suggest that public sector prices be increased. That policy, however, raises specific questions. What will the price increases have on GDP and the level of economic activities? Are the potential gains in terms of improved public finances and allocative efficiency worth the adjustment costs? How will wages and employment respond? What will be the effects on the external accounts and on transactions with the rest of the world?

The following section addresses the short-run (within a year) macroeconomic response of the Egyptian economy to a package of policy measures that includes increases in the prices of all public sector production output, as well as aggregate demand policies aimed at smoothing the disabsorption effects of the price increases. In this time frame production capacities do not change, nor do urban wages vary in response to changes in the levels of activities and allocation signals ensuing from the application of the policy package. The specific envisaged complementary policies are: (1) additional investment expenditure by state enterprises; (2) a devaluation of the commercial banks exchange rates; (3) additional foreign borrowing by the public sector; and (4) less borrowing of private sector funds by the public sector. The increased investment expenditures and foreign borrowing on the one hand and lower domestic borrowing on the other aim at maintaining the level of aggregate demand in the economy. The purpose of the devaluation is to compensate for the negative effect of the price increasers on exports going through the commercial bank pool.

The definition of macroeconomic equilibrium in MISR2 and the financing of the fiscal gap are discussed in Section III.1. It is an important mechanism embedded in MISR2 that underlies many of the results. Section III.2 provides details on the effects of increases in each public sector price, while the policy and world environment of the economy remain constant. The Section also looks at the short-run macroeconomic and distortion effects. Section III.3 considers the role of the aggregate demand policies. Finally, in Section III.4 the price increases and aggregate demand policies are combined into a package and the short-run response of the economy to this package is then analyzed.

III. 1 MACROECONOMIC EQUILIBRIUM AND THE FINANCING OF THE FISCAL GAP

A predominant outcome of the increases in public sector prices is a contraction of the economy. It can best be understood by considering the conditions of macroeconomic equilibrium summarized in Table 3.1. Relation A determines the total resources mobilized by the private sector. The latter are allocated, in relations B and C, among resources left to the private sector for its own investment expenditures (in proportion $1 - \bar{\mu}$). The parameter $\bar{\mu}$ is considered to be policy instrument in the hands of the government.[8]/ Relation D defines the fiscal gap as the resources needed by the public sector, in addition to its own savings, to carry out its

---

8/ For example, one policy may be to force banks to hold a share of their deposits in govenment bonds.

Table 3.1 - <u>Macroeconomic Balance in MISR2</u>

A. Households savings + Private company savings + Parallel market exchange rate × Net capital inflows through the parallel market = Total private sector resources

B. Net borrowing of the public sector from the private sector = $\bar{\mu}$ × Total private sector resources

C. Investment expenditures by the private sector = $(1-\bar{\mu})$ × Total private sector resources, where $0 < \bar{\mu} < 1$

D. Public Company savings + Savings of EGPC + Government savings + Fiscal gap = Public Sector investment expenditure

E. Net borrowing of the public sector from the private sector + Social security savings + Commercial banks exchange rate × Net capital inflow through commercial banks + Central bank exchange rate × Net capital inflow through central bank = Fiscal gap

investment expenditures. Finally, relation E indicates how the fiscal gap is financed: (1) borrowing from the private sector; (2) social security savings; and (3) borrowing from abroad through both the commercial bank and central bank pools. In this macroeconomic equilibrium, reflecting a realistic perception of the workings of the Egyptian economy, the following are maintained as exogenous policy variables: (1) public sector investment expenditure; (2) borrowing from abroad through the three pools; (3) the central bank and commercial bank exchange rates; and (4) the share $\bar{\mu}$ of private resources channeled to the public sector.

Consider, now, an increase in the output prices of the public sector in the absence of any other intervention, and, in particular, in the absence of variations in the share of private funds to be borrowed by the public sector. The price increases tend to improve the mobilization of savings in the public sector. The fiscal gap narrows, and more funds are available to the public sector. However, the public sector does not expand either current or capital expenditures. As a consequence, it has resources in excess of planned expenditures. If the latter are not changed, if no reduction in foreign borrowing takes place and if the share of private funds directed to the public sector does not vary, the only channel left to reaching macroeconomic balance is to reduce incomes. The latter, coupled with the cost-push elements associated with the price increases, contracts the economy.

The contraction is likely to affect the private sector more. Indeed, that sector now faces increased competition from the public sector in the markets for outputs, leading to a relative decline in private sector prices. At the same time, the cost-push effects develop. On the one hand, the increased public sector prices feed into the cost of intermediates, while on the other, inasmuch as the public sector is producing tradables, exports are affected adversely. In turn, foreign exchange and imports become more expensive, adding to the costs of intermediates. The public sector bears the burden of the higher costs of intermediates, but it is protected by the regulated prices of its output. Thus, the domestic terms of trade shift against the private sector. Because it has greater flexibility in decision-making, private sector cuts back its output.

Table 3.2 illustrates the different responses of the private and public sectors to price increases in the latter. Except for food processing, the mobilization of savings in the private sector suffers, implying less investment.9/ To the contrary, the public sector improves its resource mobilization, and its investment as a share of GDP increases.

Thus, when no other policy intervention takes place, an increase in public sector prices tends to contract the economy, a contraction that is felt more in the private sector. The reason is that price increases embody two main features: (1) an excess supply of resources in the public sector, which causes a contraction of income in the economy; and (22) the prices of private seactor output prices fall relatively, leaving the private sector facing increased costs for imports and public sector goods.

A more specific discussion of the effects of raising public sector prices is presented below, in which the various public sectors are considered individually. Following that, the effect of other policies that would eventually be associated with price increases are considered. Finally, the way these associated policies modify the effects of the price increases is analyzed.

9/ Food processing is discussed below. In essence, the price increase raises output in public sector food processing, which then raises private agricultureal output.

Table 3.2 - EFFECTS OF INCREASES IN PUBLIC SECTOR PRICES ON THE MACROECONOMIC BALANCE a/

	Base Shares in GDP	(1) Agri- culture	(2) Food processing	(3) Tex- tiles	(4) Other industry	(5) Electri- city	(6) Construc- tion	(7) Oil	(8) Transp. & comm.	(9) Services	(1) - (19) Combined
Total investment	.275	.003	-.013	.008	.047	.005	.056	.024	.020	.040	.190
Public sector	.188	.006	-.016	.017	.097	.011	.099	.054	.053	.086	.407
Private sector	.087	-.004	-.009	-.010	-.060	-.007	-.038	-.041	-.052	-.059	-.280
National savings	.216	.002	-.013	.006	.031	.004	.039	.017	.010	.028	.124
Households	.121	-.002	.011	-.005	-.019	-.004	-.008	-.030	-.016	-.034	-.107
Private	.033	-.013	-.080	-.035	-.250	-.017	-.194	-.090	-.195	-.165	-1.039
Public companies	.067	-.013	-.027	-.039	-.163	-.026	-.141	-.033	-.205	-.191	-.718
EGPC	.004	.004	-.032	.012	-.001	.005	.041	.303	-.047	.112	.397
Government savings	-.057	-.000	-.020	.005	-.061	.003	-.055	-.216	.099	.006	-.239
Social security	.048	.003	-.015	.014	.058	.007	.064	-.024	.056	.055	.218
Foreign savings	.059	.006	-.014	.019	.109	.011	.019	.052	.055	.086	.343
Domestic savings	.178	.000	-.023	-.002	-.046	.003	-.082	.040	-.005	.022	-.093
Resource gap	.097	.006	.004	.027	.219	.010	.309	-.005	.064	.074	.708
Public sector borrowing from the private sector	.071	-.004	-.009	-.010	-.060	-.007	-.038	-.041	-.052	-.059	-.280

a/ The percentage change in the indicators for a 1 percent in price. The elasticities are obtained from a perturbation of the 1982/83 solution.

III.2 Increasing Public Sector Prices

Public sector activities are not homogenous. Some produce tradables, others non-tradables, some sectors face a regulated price but can determine their level of activity and let demand adjust, others have to satisfy demand at the regulated price. To account for these differences and the implications of price adjustments, public sector activities are classified into four categories: (1) supply-driven sectors producing tradables; (2) supply-driven sectors producing non-tradables; (3) demand-driven sectors; and (4) oil. In the supply-driven sectors, producers maximize their profits at the regulated price; thus they determine their level of activity, and the domestic market for their output is rationed. In the demand-driven sectors, supply is perfectly elastic, and the level of activity can be below or above the maximum profit level. In the latter case, producers incur a negative rent.

2.a Supply-Driven Sectors Producing Tradables

Agriculture, food processing, other industry, and transportation and communications are supply-driven sectors producing tradables. They maximize their profits at a regulated price, which determines their level of supply. Given the price of substitutable commodities on the world market and the specific exchange rate, the demand for exports takes its share of the supply, as do, eventually, the government-trading agencies. The remaining supply is for the domestic market, which is thus rationed. Public sector activities have a fixed wage bill and can increase production only through more intensive use of intermediaries. Tables 3.3 and 3.4, columns

1, 2, 3, 4 and 8, show the percentage change in the macroeconomic and distortion indicators that results from a 1 percent increase in prices in the supply-driven sectors producing tradables.

Agriculture

Public sector agriculture is relatively small, with little weight in the economy. This is shown by the minor changes induced by an increase in the price of the output of this sector (see Table 3.3), column 1). The general effect is a slight contraction of the economy, with a depreciation of the parallel market exchange rate, a decline in the rural wage and an improvement in the finances of the public sector. The implications for distortions are mixed: they are alleviated in some areas and exacerbated in others.

The contraction of the economy comes about as follows. The increase in the output price hurts agricultural exports, which are channeled through the central bank pool of foreign exchange. This adverse effect leads to a larger transfer of foreign exchange from the commercial bank pool to the central bank. The availability of foreign exchange in the commercial bank pool is consequently reduced implying a tighter foreign exchange budget and a reduction in the quantity of imports possible through the commercial bank pool. The demand for imports then tends to spillover to the parallel market, inducing a depreciation in the exchange rate there. That depreciation in turn tends to lower the share of workers' remittances coming into the country through the commercial bank pool, further exacerbating the shortage of foreign exchange in that pool. The result is more expensive

Table 3.3 - EFFECTS OF ADJUSTMENTS IN PUBLIC SECTOR PRICES—MACROECONOMIC AGGREGATES [a]

	(1) Agri-culture	(2) Food processing	(3) Tex-tiles	(4) Other industry	(5) Electri-city	(6) Construc-tion	(7) Oil	(8) Transp. & comm.	(9) Services	(1) - (19) Combined
Selected constant price aggregates										
GDP	-.006	-.036	-.035	-.069	-.003	-.001	+.003	-.072	-.044	-.262
Private consumption	-.006	-.028	-.031	-.077	-.044	-.096	-.037	-.042	-.057	-.378
Public consumption	.000	-.029	-.004	+.006	-.007	+.026	+.019	+.003	+.012	+.026
Public investment	-.001	-.029	-.005	-.048	+.008	+.130	+.074	-.035	+.042	+.136
Private investment	-.007	-.011	-.025	-.124	-.004	+.269	+.005	-.131	-.027	-.055
Total investment	-.003	-.023	-.012	-.074	+.004	+.178	+.050	-.068	+.001	+.053
Imports GNFS	-.004	-.005	-.008	-.032	-.001	+.026	+.074	-.045	-.031	-.026
Exports GNFS	-.008	-.032	-.063	-.054	-.001	+.057	+.141	-.182	-.095	-.237
Other macroeconomic variables										
Current account deficit	+.000	+.002	+.001	+.012	.000	+.020	.002	+.002	+.000	+.035
"Parallel exchange rate"	+.004	+.023	+.019	+.181	.000	+.298	.032	+.028	+.002	+.523
Urban wage	.000	.000	.000	.000	.000	.000	.000	.000	+.000	+.000
Rural wage	-.006	+.262	-.005	-.022	-.024	-.147	-.087	+.013	-.179	-.195
Urban CPI	-.001	+.042	+.018	-.007	-.009	+.028	-.057	-.040	-.042	-.068
Rural CPI	-.003	+.068	+.008	-.037	-.015	-.042	-.078	+.003	-.110	-.206
Employment	-.000	+.002	+.004	+.002	-.002	+.001	-.009	-.013	-.003	-.018
Selected public finance indicators										
Government revenue	+.001	+.011	+.004	+.028	.000	+.023	+.051	-.021	+.003	.100
Indirect taxes	.000	+.050	+.005	+.117	-.004	+.093	+.087	-.090	-.064	.194
Direct taxes	.000	-.012	+.002	-.040	+.001	-.034	-.007	+.016	+.017	-.057
Subsidies	-.001	+.043	+.009	-.001	-.005	-.023	+.010	-.054	-.047	-.069
Fiscal gap	-.005	+.003	-.012	-.075	-.008	-.066	-.060	-.042	-.068	-.333

[a] The percentage change in the indicators for 1 percent in price. A is derived from a perturbation of the solution in 1983/84, while B and C are the outcome of perturbations of the solution in 1982/83.

Table 3.4 - EFFECTS OF ADJUSTMENTS IN PUBLIC SECTOR PRICES ON DISTORTIONS a/

	(1) Agriculture	(2) Food processing	(3) Textiles	(4) Other industry	(5) Electricity	(6) Construction	(7) Oil	(8) Transp. & comm.	(9) Services	(1) - (19) Combined
Shares of Rents in GDP										
Subsidized consumer goods	-.020	-.146	-.084	-.270	.023	-.103	-.182	-.042	-.047	-.917
Rents accruing to urban households	-.020	-.084	-.270	-.023	-.103	-.182	-.042	-.047	-.047	-.917
Rents accruing to rural households	-.037	+.124	-.192	-.681	-.051	-.625	-.270	-.441	-.469	-2.64
Total rents to households	-.032	+.047	-.161	-.564	-.043	-.476	-.245	-.327	-.348	-.215
Public sector employment	-.183	-.665	-.572	-.255	-.166	-.713	-.064	-.968	+1.14	-4.74
Public sector output markets										
Supply-driven sectors	-.096	-.735	-.371	-2.01	-.092	-1.46	-.557	-1.08	-1.12	-7.52
Demand-driven sectors	+.063	+.652	+.241	+1.473	-1.07	-.092	+.102	-.866	-6.39	-5.89
Imports premia	+.041	+.117	+.043	+.997	+.005	+1.18	-.263	+.416	+.170	2.71

a/ The percentage change in the indicators for a 1 percent change in the prices of the sector. The elasticities are obtained from a perturbation of the 1982/83 solution.

imports at lower quantities. The increased cost of imports adversely affects production in the sectors where there is intensive use of imported inermediates and where substitution possibilities are limied. At the same time, the price increase in agriculture affords a larger supply to the domestic market through both greater output and lower exports. Public agriculture therefore competes with private agriculture, exerting a downward pressure on prices. This latter effect, jointly with the cost increases, hurts the private sector and leads to a contraction in income.

The improvement in the finances of the public sector is the outcome of lower subsidies and increased revenues. Two categories of subsidies are distinguished in MISR2: those generated by the trading activities of government agencies, and those directly linked to the level of activity in various sectors. The contraction in the economy contributes to a reduction in the value of the subsidies as follows: (1) the government trading agencies face relatively lower prices for output from private sector food processing, which reduces their outlays; and (2) other subsidies decline as a result of the lower level of activity in the economy. Higher government revenue results mainly from the improved profitability of the agricultural sector, induced by the price increase. Indeed, raising the output price causes a notable outward shift in the marginal productivity schedules of the fixed factors, labor capital and land. For the latter two, the implication is a direct increase in income whereas for labor the price increase reduces the negative rent imposed by the fixed wage bill.10/

Consider now the effects of the price increase on distortions. Table 3.3, column 1 presents the elasticities of the ratio of rents to GDP.

10/ In addition, higher import premia feed public sector revenues.

Government agencies supply households with certain amounts of agricultural and industrial goods as well as processed food at subsidized prices. The quantities available are represented in Figure 2.1B as q_d; they are supplied at the subsidized price p. Given their demand schedule at D, households will be willing to pay the virtual price P_v. The adverse effects of the increase in the output price of public sector agriculture cause curve D to shift inwards, reducing the virtual price and the implied rent. The fall in the latter is larger than that in GDP, leading to an alleviation of the distortions in the markets considered. Similarly, as mentioned earlier, the increase in the output price shifts the marginal products of labor in public agriculture outwards. Such a shift closes the gap between the virtual and actual wage, reducing the level of distortion. The reduction in the distortion in the markets faced by the supply-driven sectors also results from the contraction of income in the economy, which shifts domestic demand inwards, again reducing the gap between virtual and market prices.

In the case of the demand-driven sectors, the distortions are exacerbated. These sectors essentially produce non-tradables whose relative prices are now lower after the induced increase in the cost of imports. Consequently the sectors face additional demand and their output expands. The expansion widens the gap between their marginal costs and the regulated price they face, leading to more distortions. Finally, raising the output price in public agriculture increases the rents associated with the rationing of imports. Indeed, foreign exchange in both the central and commercial bank pools becomes less available, contributing to more rationing and to a larger wedge between the opportunity cost of imports and their actual price.

Food Processing

Consider now an increase in the output price of public sector food processing. It has the same overall contractionary effect as does raising the price of agriculture: GDP and all its components are adversely affected (see Table 3.3, column 2). However, in contrast with agriculture, and despite of an increase in government revenues, the fiscal gap widens. Similarly, the rural real wage, urban and rural consumer price indexes (CPIs) and employment increase. The results on distortions are mostly similar to the ones produced by the price increase in agriculture, except in the case of the rent accruing to rural households, which increases as a share of GDP.

The contraction of the economy comes about in the same way as previously, but with some minor modifications. Exports of processed food are assumed to go through the parallel market. Hence, in the first round the availability of foreign exchange in either the central or commercial bank pools is not affected. The parallel market exchange rate depreciates, leading in subsequent rounds to a decline in the share of workers' remittances going to the commercial bank pool. As a consequence, imports in that pool are cut back. Again, because imports are more expensive and less available, marginal costs and production are adversely affected. In the case of public processed food, however, unlike with agriculture, the higher cost of imports is not sufficient to wipe out the incentive resulting from the price increase. Production expands, generating more demand for domestic intermediates, which come from private agriculture. The expansion of the latter sector pulls rural wages up. Because of trade margins and the non-tradable nature of their output, the public and private service sectors expand, pulling more labor into the economy. One outcome of the increase in the price of public sector ood processing is thus to shift the distribution of income in favor of rural areas.

One purpose of the increases in public sector prices is to improve public finances. This outcome does not seem to result from raising the price of food processing. Indeed, direct taxes fall and subsidies increase. The decline in direct taxes is induced by the shift of incomes toward rural households, whose effective tax rate is lower. Similarly, the relative increase in the cost of processed food in the economy while maintaining the same subsidized price adds to the burden of government trading and pushes up the value of subsidies. Overall, in spite of the increase in indirect taxes, the fiscal gap widens.

In terms of distortions (see Table 3.4), column 2), the outcome of raising the price of food processing is similar to that form agriculture, with one exception: the share in GDP of the rent accruing to rural households increases. The expansion of public sector food processing boosts private agriculture, leading to increased real rural income. The consequence is an outward shift in rural household demand which raises the virtual price of the rationed goods available to them.

Textiles

An increase in the output prices of the textiles sector has the same contractionary effect as an increase in the price of food processing (Table 3.3, column 3). The parallel market exchange rate depreciates, through to a lesser degree and employment expands. The real rural wage, howver, declines, and the fiscal gap improves. Another difference with food processing is found in the decline of the share in GDP of rents accruing to rural households. Unlike food processing, but as with all other sectors, textiles have a weak link with private agriculture and hence do not exert the pull required to increase the real rural wage. The improvement in the fiscal gap is mainly the outcome of the improved profitability of the textile sector and lower increase in subsidies. Indeed, the increase in the price of textiles output, as in the case of the other sectors, improves the marginal productivities of the fixed factors, labor and capital. With the former, this improvement is translated into a lower negiative rent as a result of the fixed wage bill; in the case of capital, it implies a higher rate of return. Together these two elements enlarge the savings of public companies and reduce the fiscal gap. Finally, because rural wages and income are not expanding, there is less rural consumption on the rationed goods. Their virtual price declines, and the level of distortion introduced by the rationing of households is diminished (see Table 3.4, column 3).

Other Industry

By comparison with other sectors, an increase in the prices of other industry sector produces much larger effects (see Table 3.3 and 3.4, columns 4). They are, however, mainly in the same direction: a contraction of the economy, depreciation of the parallel market exchange rate, and improvement in the fiscal gap. This price increase also produces a relative price change that makes the bundle of goods purchased by the government relatively less expensive. Hence, there is a slight increase in real public consumption. In terms of distortions, the same pattern as for textiles is obtained: distortions are alleviated everywhere except in the markets for the demand-driven sectors and for imports.

Transportation and Communications

The macroeconomic response of the economy to an increase in the price of transportation and communications follows the same pattern as the one obtained with other industry (see Table 3.3, column 8). One difference, however, is that the exports of transportation and communications go through the commercial bank pool of foreign exchange. The increase in the output price of the sector has therefore a stronger negative impact on exports and imports. Indeed, the fixed commercial bank exchange rate does not have the dampening effect on the decline of exports that the flexible parallel market exchange rate has on the exports of the three industrial sectors. Similarly, the resulting lower availability of foreign exchange in the commercial bank pool leads to a more important cut in imports. Another difference with the commodity-producing sectors is the decline in the revenues of the government. This is caused mainly by a decline in the collection of indirect taxes. Two reasons underlie the decline: (1) the general contraction of the economy; and (2) on the one hand, the low share of indirect taxes on services in total indirect taxes and, on the other hand, the induced relative expansion of the service sectors. Finally, because of the increase in the cost of imports relative to that of domestic goods, and because of the higher domestic intensity of the intermediates used in the service sector, there is a relative decline in the marginal cost of the latter. In spite of the expansion of the public service sector, that decline alleviates the distortions for the demand-driven sectors (see Table 3.4, column 8).

2.b Supply-Driven Sectors Producing Non-Tradables

One essential feature of the construction sector is that its output is non-tradable. It is assumed in MISR2 to be facing a regulated price and to maximize its profits at that price. There is no demand for exports, and domestic users are rationed. Raising the price of construction produces a significantly different result from increases in the prices of other sectors: the contraction of the economy is almost negligible and is only the result of a decline in private consumption (see Table 3.3, column 6). Most other results, however, are similar to the ones obtained by raising the regulated prices of the other supply-driven sectors. The finances of the public sector improve; the parallel market exchange rate depreciates. All distortions except those attributable to the rationing of imports are alleviated. The increase in the price of public sector construction does not directly affect exports as the sector produces non-tradables.

Two effects that work in opposite directions are generated. On the one hand, an expansion in the output of construction drives its relative cost down. On the other hand, the expansion means a larger demand for intermediates, which pulls up some prices and contributes to the depreciation of the parallel market exchange rate through more imports. The cost-push elements reduce output in some sectors, while the demand-pull elements expand output in others. The private sectors except other industry are more affected by the cost-push developments: the contraction of their output affects private income and consumption, which decline. Apart from construction and the demand-driven sectors, the levels of activity of all other public sectors depend on variations in their costs. Only food processing and transportation and communications see a decline in their relative costs and an expansion in their output. On balance, however, the savings of the public companies improve. Because the relative costs of public consumption and public investment decline, and because they are budgeted for in nominal terms, their real values increase. The expansionary factors are sufficient to pull more imports into the economy, causing a depreciation in the parallel market exchange rate. By adding more resources in domestic currency to the pool of private savings, the depreciation, together with the relatively lower cost of investment goods, contributes to an increase in real private investment. Simultaneously, the depreciation improves the competitiveness of the economy and exports of non-agricultural commodities expand. More oil is also exported because of the contraction of private consumption.

The revenue of the government grows, mainly because it collects more tariffs on imports, which contribute to the expansion of indirect taxes. A less significant addition to government revenue comes from the improved profitability of public sector construction. The slowdown of the private economy adversely affects direct taxes. Simultaneously, the lower price of private food processing diminishes the cost to the government trading agencies, and subsidies decline. Overall, as for all other public sector production activities except food processing, the fiscal gap is reduced.

Naturally, the contraction of private consumption reduces the relative virtual prices of rationed goods. It reduces the wedge between the opportunity cost and the market price of the specific goods and consequently alleviates distortions there (see Table 3.4, column 6). With respect to employment, the increase in the price of public construction improves the marginal productivity of labor in the sector and diminishes the level of distortion. The improvement is sufficient to reduce the ratio to GDP of the negative rents imposed on the public sector by the fixed wage bill. By expanding supply, the increase in the price of construction reduces its scarcity value and closes the gap between its opportunity cost and market value; the supply-driven sectors on the whole face less distortion. By contrast with the result obtained for the other price increases, the demand-driven sectors are also favorably affected by the higher price of construction. The demand they face contracts more than the expansion of their costs; the gap between their marginal cost and their regulated prices closes. Finally, the additional demand for imports exacerbates their scarcity, and the premia associated with imports increase.

2.c The Demand-Driven Sectors

Electricity and services are assumed to be demand-driven: they face a regulated price and their supply adjusts to meet demand. If the latter happens to be beyond the amount that would allow them to maximize their profits, they incur a negative rent; if, on the contrary, demand is below the maximum profit level of output, they receive a pure rent.

Electricity

The electricity sector produces a non-tradable. Because of the relatively small level of the output of electricity, raising its price has relatively minor effects on the rest of the economy (see Tables 3.3 and 3.9, columns 5). However, it does depress the demand for electricity and its level of activity. It also contributes to a slightly increased domestic demand for oil, which adversely affects the exports of the central bank pool. The additional foreign exchange required in that pool needs to be bought from the commercial bank pool, leading to a tighter foreign exchange budget there. This latter effect, combined with the general contraction of the economy, results in a drop in imports. However, relative price changes are such as to reduce the cost of the bundle of goods required for public investment, thus contributing to a real increase in public sector capital formation.

As with the price increases in most other sectors, the fiscal gap narrows. On the revenue side, the main sources of improvement are the higher direct taxes and savings of public companies. The better performance of these two resides mainly in the increased profitability of electricity, which improves the income of public companies. On the expenditure side, the price of electricity has a depressing effect that helps reduce the level of subsidies. On the one hand, the subsidies linked to the level of activity of various sectors decline, and, on the other, the relatively lower price of private food processing reduces the cost to the government of food subsidies. Overall the fiscal gap narrows.

The effects on the distortions are as expected. All of them except for those associated with imports are alleviated. The rents accruing to households fall because of reduced income; the negative rents caused by the fixed wage bills shrink, following the higher marginal productivity of labor in electricity; the contraction of the economy diminishes the scarcity value of the goods produced by the supply-driven sectors; while the reduced wedge between the marginal cost and market price of electricity lowers the level of distortion for the demand-driven sectors. However, the slight reduction in exports, by reducing the availability of foreign exchange and forcing a cut in imports, contributes to higher import premia.

Services

The other demand-driven sector is services. The same pattern of results is produced by raising the price of electricity, although the effects are more important because the sector is larger (see Tables 3.3 and 3.4, columns 9). Another difference with the electricity sector is that a small amount of the output of services is exported. Thus an increase in the price of services dampens both domestic and foreign demand. The reduced exports of services, which go through the commercial bank pool, reduce the foreign exchange there. Imports financed via the commercial banks are therefore cut, and some imports spill over to the parallel market, leading to the depreciation of the exchange rate, which helps, albeit not sufficiently, exports of non-agricultural commodities. On balance, the outcome is fewer and more costly imports, as well as less exports. These latter two effects contribute to a slackening of the economy. Thus, there is a reduction in absorption and a switch from foreign to domestic demand. However, the switch is not sufficient to compensate for the cut in absorption. The end result is a reduction in private sector activity except for private services. The latter receive some of the demand switched out of the public sector. At the same time, some public sector activities emerge with better relative prices and expand their production, although only to a limited extent. As a result of the relative price changes, the costs of public consumption and investment are reduced, and these two components of aggregate demand expand in real terms.

Public finances improve in the same way as in the other sectors. The cost of subsidized goods is relatively lower through the depressed price of private food processing; the savings of public companies and the direct taxes they pay are larger because of greater profitability. The overall fiscal gap is reduced.

The results of the increase in the price of services is similar in terms of distortions to the one obtained for electricity and for the same reasons. However, the distortions in the output market in the demand-driven sectors are reduced substantially as the gap between marginal cost and regulated price is reduced by quite a lot.

2.d Oil

Rarely has the Egyptian economy traditionally resource-constrained, had an opportunity such as the one oil is providing. The problem is to exploit the opportunity in the most favorable way. The long-run issues of the investment strategy posed by oil are treated very well in Dervis, et al., (1983). In the short- to medium-run, the question is the pricing of oil and the effect of bringing the domestic price to the world level. The implication of raising the domestic price within a year is considered here (see Tables 3.3 and 3.4, columns 7).

Unlike with all the other sectors, raising the price of oil increases GDP at constant market prices. The only negative effect is borne by private consumption, and it is not substantial. The primary effect of raising the price of oil is to expand oil exports through a cut in domestic consumption. The additional foreign exchange available to the central bank pool, through which oil is exported, releases more foreign exchange to the commercial banks. In turn, the larger availability of foreign exchange in the latter pool allows for more imports and less spillover to the parallel market. As a result, the parallel exchange rate appreciates. The outcome is more and less costly imports. Simultaneously, the appreciation of the parallel rate hurts exports of non-agricultural commodities going through that pool, where transactions are governed by that rate. The slack in the export demand on the one hand expands the supply from the public sector to the domestic market, and on the other depresses the demand faced by the private sector. The adverse shift in the private sector demand results in lower relative prices there. Given the appreciation of the exchange rate, the lowering of prices in the private sector reduces the relative cost of investment goods and the bundle of commodities for public consumption. These components of aggregate demand increase in real terms and more than compensate for the decline in private consumption.

The expected positive effect of the increase in oil prices on public finances is obtained. The main reason is the improved performance of the oil sector, which generates more operating surplus, contributing to more public sector savings. However, at the same time the contraction of the private economy adjusting to the increased costs of energy leads to a decline in direct taxes. However, indirect taxes, following the expansion of all the non-private consumption components of GDP, increase. For the same reason, the level of subsidies is also higher. Overall the improved operating surplus in the oil sector is the dominant effect, leading to a reduction in the fiscal gap.

The decline in private income and consumption underlies the lower shares of rents accruing to households because of the subsidization of certain goods. Indeed, the reduced income leads to an inward shift in the demand curves, which produces relatively lower virtual prices: the level of distortion is reduced. As with the price increases in other sectors, the distortions resulting from the fixed wage bill are alleviated, although through different channels. The increase in the price of oil adversely affects the private sector and leads to a relative drop in output prices there. This latter effect, jointly with the appreciation of the parallel market exchange rate and the fixed prices in the public sector, implies lower marginal costs and higher levels of output in the public sector. As a consequence, the scarcity value of labor rises implying less distortion. By the same mechanism, the expansion of activity in the public sector relaxes the rationing of domestic markets in the supply-driven sectors, implying a reduction in the level of distortion there as well. By contrast, however, the gap between opportunity cost and market prices in the demand-driven sectors widens. Indeed, the expansion of the public commodity sectors implies, through trade margins, a larger demand and output in public sector services. That output is supplied at the same regulated price but at a higher marginal cost. Finally, as can be expected, the share of import premia in GDP declines: the larger amounts of oil exported make more foreign exchange available and relax the import budget through the commercial bank pool.

III.3 Aggregate Demand Policies

The general conclusion emanating from this discussion is that an increase in public sector prices tends, in the short run, to contract the economy. This conclusion is, however, based on the premise that no other policy is implemented simultaneously: public prices are increased without the government changing its policies on domestic and foreign borrowing. If a contraction of the economy were anticipated, it is likely the government would implement, jointly with the price increases, polices aimed at containing the contraction, if not at maintaining the level of activity.

Four policies directed toward such an outcome are examined here. In the first, public companies increase their level of investment, while in the second, a lower share of private funds is channeled to the public sector. In other words, the public sector borrows less from the private sector. The two other policies are related to transactions with the rest of the world. One calls for more foreign borrowing through the commercial bank pool, with the purpose of alleviating the burden of the increased public sector investment outlays on domestic funds. The last policy is a devaluation of the commercial bank exchange rate in order to contain the adverse effect of the price increases on the exports passing through this pool. These four policies are considered first independently and then in combination.

3.a Public Investment Outlays

Columns 1 of Tables 3.5 and 3.6 give selected indicators measuring the effect of a 1 percent increase in investment by the public sector. GDP at constant market prices increases, as do private consumption and exports. Real public sector investment shows a relatively minor gain. All other components of real GDP are adversely affected. The level of employment improves while the parallel market exchange rate depreciates. In spite of a nominal increase in the rural wage, it declines in real terms. The fiscal gap worsens. Distortions are exacerbated. The increased public sector investment has two effects, higher aggregate demand and excess demand for funds.

The higher aggregate demand can only be accommodated through supplies from the private sector, the demand-driven public sectors and imports. The other public sectors, which are supply-driven, cannot expand their supply. The larger demand faced by the private sectors directly, or indirectly through the expansion of the demand-driven public sectors, puts pressure on private sector prices, which show a relative increase. Similarly, the initial larger demand for imports leads to a depreciation in the parallel market exchange rate. That latter effect partially shifts workers' remittances out of the commercial bank pool and increases the costs of imports. These two effects dampen the demand for imports, which ultimately decline. However, the depreciation has the advantage of helping non-agricultural commodity exports, which pass through the parallel market. The relatively higher costs of imports and private sector goods have an adverse effect on the public sector, which cannot adjust its output prices.

Table 3.5 - Effects of Policies Aimed at Maintaining the Level of Aborption a/

	(1) Increase in public company investment expenditure	(2) Lower share of public borrowing from the private sector	(3) Increase in net foreign borrowing through the commercial bank pool	(4) Devaluation of the commercial bank exchange rate	Combined policies 1+2+3+4
A. Selected constant price aggregates					
GDP	.004	.023	.018	.095	.140
Private consumption	.011	.079	.019	.070	.179
Public consumption	-.033	-.123	.093	.036	-.027
Public investment	.029	-.315	.149	.103	-.034
Private investment	-.044	.615	.072	.190	.833
Total investment	.006	-.016	.122	.133	.245
Imports GNFS	-.003	-.013	.121	.195	.300
Exports GNFS	.005	.003	-.031	.314	.291
B. Other macroeconomic variables					
Current account deficit	.012	.002	.608	.607	1.23
Parallel exchange rate	.179	.038	-.099	-.124	-.006
Urban wage	.000	.000	.000	.000	.000
Rural wage	.185	.176	.015	.254	.630
Urban CPI	.231	.183	-.089	.040	.365
Rural CPI	.266	.194	-.074	.058	.444
Employment	.027	.021	-.003	.045	.090
C. Selected public finance indicators					
Government revenue	.184	.101	-.055	-.071	.159
Indirect taxes	.118	.059	-.009	.122	.290
Direct taxes	.271	.164	-.092	-.128	.215
Subsidies	.095	.058	-.007	.090	.236
Fiscal gap	.237	-.221	.152	.265	.433

a/ The figures indicate the percentage change in he indicators for a 1 percent change in the indicastor. A is derived from a perturbation of the solution in 1983/84, while B and C are the outcome of perturbations of the solution in 1982/83.

Table 3.6 - Effects of Maintaining the Level of Absorption on Measures of Distortion a/

	(1) Increase in public company investment expenditure	(2) Lower share of public borrowing from the private sector	(3) Increase in net foreign borrowing through the commercial bank pool	(4) Devaluation of the commercial bank exchange rate	Combined policies 1+2+3+4
Share of rents in GDP					
A. Subsidized consumer goods					
Rents accruing to urban households	-.528	1.031	-.095	.360	1.89
Rents accruing to rural households	.850	.408	-.056	.550	2.12
Total rents to households	.758	.586	-.067	.497	2.05
B. Public sector employment	.313	.014	-.251	-1.34	-2.02
C. Public sector output markets					
Supply-driven sectors	2.35	1.18	-.510	.922	3.53
Demand-driven sectors	.150	.391	.073	2.66	3.62
D. Import premia	.592	.223	-.914	-4.11	-4.38

a/ The figures indicate the percentage change in the indicator for a 1 percent change in the prices of the sectors. The elasticities are obtained from a perturbation of the 1982/83 solution.

adjust its output prices. Besides the level of activity in the public sector, the components of GDP defined in nominal terms show a decline. The increase in investment by the public sector allows it to keep ahead of the price increases and produces a relatively weak expansion in real public investment.

The other aspect of greater investment by the public sector is an excess demand for funds at the macroeconomic level. The gap leads to three types of adjustments. The initial increase in aggregate demand generates more income by expanding essentially private sector activities. Simultaneously, the depreciation of the parallel market exchange rate increases the domestic currency value of foreign savings. The latter two effects, by raising prices in the domestic economy, cut the real value of all nominally fixed magnitudes and, in particular, public sector investment expenditures. Thus the economy tries to accommodate the additional investment expenditure by increasing the level of activity. However, it rapidly faces resource constraints that raise private sector prices and the cost of foreign goods. The expansion becomes more and more costly. In the process the real value of nominally fixed magnitudes declines.

The changes in the distortion indicators shown in column 1 of Table 3.6 follow from the expansion of the private economy, the contraction of the public sector and the deterioration in the foreign accounts. The increased share in GDP of the rents accruing to households is based on the expansion of their income. Indeed, the expansion in income leads to outward shifts in household demand that generate a relative increase in virtual prices. Hence, the gap between the opportunity cost and subsidized value of the rationed goods grows larger. The greater negative rent attributable to the fixed wage bill in the public sector results from the contraction of output there. The latter causes inward shifts in the marginal productivity of labor schedules, driving virtual wages down and further away from the regulated wages. The substantially higher rents occurring in the markets of goods produced by the supply-driven sectors result from both the expansion of aggregate demand and the contraction of public sector outputs. Indeed, these two developments exacerbate the scarcity of the goods produced by these sectors and hence increase their virtual costs. The larger share in GDP of the negative rents generated in the markets of the outputs of the demand-driven sectors are the result of the increases in their marginal costs. Finally, the tighter availability of foreign exchange in the commercial bank pool implies a more binding rationing of imports that increases their scarcity value.

On balance, in the absence of other policies, an expansion of investment by the public sector leads to higher GDP at constant market prices. This outcome results from simultaneously higher and lower levels of activity in the private and public sectors, respectively. Hence, in the short run, the additional investment barely increases. Public sector consumption in real terms declines.

3.b Less borrowing from the Private Sector

One important reason for raising public sector prices is to improve the mobilization of public resources. If that is the expected outcome, then there is less need for the public sector to borrow funds from the private sector. This section considers the short-run macroeconomic consequences of reducing the share of private resources channeled to the public sector, in the absence of any other intervention.

Column 2 of Table 3.5 provides the percentage changes in selected macroeconomic indicators in response to a 1 percent decline in the share of private funds channeled to the public sector. That change has two initial effects. On the one hand, more resources are available for private sector investment, and on the other there is excess demand for funds in the public sector. These two effects work in the same direction and affect the economy through the same channels as does the increase in investment by the public sector considered above. Private sector prices show a relative increase; the parallel market exchange rate depreciates. As a consequence, private sector output expands, while the level of activity in the public sector declines because of higher costs and unchanged output prices. Most indicators move in the same direction as with the increase in public sector investment expenditures. One expected difference is the increase and decline in real private and public investment respectively. The relative increase in investment prices is not sufficient to offset the higher investment expenditure by the private sector. The fixed public sector investment budget absorbs the increase in investment prices through a decline in real investment. Another result that is different by comparison with what larger public sector investment produces is the lower fiscal gap. Where there is investment, some financing is obtained from the private sector through the expansion of its resources in the latter does not offset the decline in the share of these resources channeled to the public sector. Instead, borrowing from the private sector is replaced by the resources of the public sector itself. However, this shift is obtained at the cost of relatively higher investment prices that reduce the real value of public sector investment.

The effects on distortions, as shown in column 2 of Table 3.6, are identical to the ones produced by the increase in public sector investment. The expansion of the economy in the face of limited resources and rationing exacerbates the distortions.

3.c Additional Foreign Borrowing

Raising public sector prices tends to reduce the fiscal gap, generating resources for the public sector. If nothing is done with these resources and no other policy is adopted, the economy contracts. In order to avoid the contraction and help an eventual restructuring of the public sector, the additional resources can be used to finance increased investment by public companies. However, the domestically generated resources may not be sufficient for the higher level of investment, especially if the public sector also reduces its borrowing from the private sector. A short-run solution is to raise the level of foreign borrowing to smooth the adjustment. In order to disentangle the effects of the higher foreign borrowing from those of the other elements of the package of policy measures, the response of the economy to an increase in the amount of foreign borrowing through the commercial bank pool, to the exclusion of any other policy, is considered here.

Column 3 of Table 3.5 presents the changes in selected indicators following a 1 percent increase in foreign borrowing through the commercial bank pool. GDP at constant market prices and all its components apart from exports expand. The parallel market exchange rate appreciates, while the real rural wage increases. Public sector revenues decline, and the fiscal gap widens.

If the public sector engages in more foreign borrowing, the initial consequence is an excess supply of resources at the macroeconomic level. In the absence of other interventions, the nominal level of income needs to drop. But the costs to domestic users decline even more relative to world prices, allowing for an expansion of the level of activity. The larger availability of foreign exchange in the commercial bank pool means a larger inflow of cheaper imports and less recourse to the parallel market, implying an appreciation of the exchange rate there. This effect is further accentuated by a shift of workers' remittances out of the parallel market into the commercial bank pool because of the appreciation. The latter shift and a lower cost of imports relax the cost pressures and lead to expansion of output in both the public and private sectors. The larger supplies in the economy drive prices in the private sector down implying a lower expansion of output there than in the public sector, whose prices are fixed. Now imports and private sector goods are relatively less expensive. Hence all expenditures fixed in nominal terms, in particular public consumption and investment, have a higher real value. As to outcomes, exports decline because of the appreciation of the parallel market exchange rate on the one hand and the real expansion of the economy on the other. The former has an adverse effect on the exports of non-agricultural commodities assumed to pass through the parallel market, while the expansion creates a larger domestic demand for oil and contracts the exportable surplus.

The additional foreign borrowing does not benefit domestic resource mobilization in the publc sector: the main sources of revenues decline. Moreover, despite the reduction in the value of the subsidies, this situation induces a deterioration in the fiscal gap. Direct and indirect taxes as well as most revenues of the government are pegged to nominal income. Following the increased foreign borrowing, the latter contract, decreasing public sector revenues. In particular, the public sector no longer receives the import premia, which decline as a result of the larger availability of foreign exchange and less binding rationing of imports. Given that no change occurs in the current consumption of the government and in public sector investment, the fiscal gap widens.

The short-run effects on distortions of a 1 percent increase in foreign borrowing are shown in column 3 of Table 3.6, which presents the implied percentage change in the ratio of the specific rent to GDP. Because of the positive elasticity of demand with respect to income and the lower levels of income, the ratios of household rents to GDP decline, reducing the level of distortion. On the employment side, firms in the public sector benefit from their regulated prices and the lower costs of imports and private sector goods. As a result, their output expands, contributing to relatively higher virtual wages: the negative rent attributable to the fixed wage bill declines. Similarly, distortions in the markets of the outputs of the supply-driven sectors are also alleviated because of the expansion of outputs, which reduces the level of rationing in these markets. As can be expected, the expansion of output in the demand-driven sectors enlarges the gap between their marginal cost and the price they receive, implying a wider difference between opportunity costs and market prices. Finally, the larger availability of foreign exchange, by alleviating the rationing, also reduces the share of import premia in GDP.

3.d Devaluation of the Commercial Bank Exchange Rate

The last instrument in the policy package is a devaluation of the commercial bank exchange rate. On its own, the devaluation allows an expansion of GDP and all its components at constant market prices. However, it also reduces government revenue and leads to a widening of the fiscal gap.

Because of the particular trade regime in Egypt, a devaluation leads to a relative decline in import prices relative to domestic prices. Exports passing through the commercial bank pool expand, while workers' remittances shift out of the parallel market. The additional foreign exchange allows a relaxation of the rationing of imports, which come in more through the commercial bank pool, reducing the demand for imports in the parallel market. The shift of imports between the two pools reduces the pressure on the parallel market exchange rate, which appreciates. The appreciation and reduced rationing allow a relative decline in import prices and a reduction in import costs. As a consequence, the nominally fixed expenditure increase in real terms and domestic output expands. More income is generated in the private sector and private consumption increases. The latter benefits also from the higher value in domestic currency of the transfers from abroad. Rates of return in the public sector also improve. Following the upswing in aggregate demand, private sector prices rise.

Another aspect of the devaluation is the larger amounts of resources available to finance public sector investment. However, lower income in the public sector reduces this excess supply of resources, in spite of the higher public rates of return and the expansion of the economy. This latter result happens in two ways. On the one hand, the lessened constraint on imports reduces import premia substantially and thus dries up an important source of revenue for public sector companies. That reduction in import premia more than offsets the improved rates of return. On the other hand, the larger aggregate demand reduces the exportable surplus of oil. The income resulting from the difference between the world and domestic prices of oil therefore falls contributing to lower public sector revenue. On balance, the fiscal gap widens.

As column 4 of Table 3.6 shows, the share in GDP of the rents accruing to households increases following the expansion of private consumption. Similarly, the higher level of activity in the economy exacerbates the distortions in the markets of both the supply-driven and demand-driven sectors. In the former, supply lags behind the larger aggregate demand, thus accentuating the scarcity of public sector goods. In the demand-driven sectors as well, demand grows faster than the decline in marginal costs. As expected, the devaluation reduces import premia, with the adverse consequence on public finances mentioned above. However, the expansion of output in the public sector narrows the gap between virtual and regulated wages.

III. 4. The Effects of the Combined Package

Policy-makers are often told that prices are out of line with opportunity costs. Most of the time prices are too low, encouraging too much demand, reducing efficiency and hurting resource mobilization. The recommended solution is always to raise the uneconomic prices to the level of their opportunity costs. The advisers are rarely concerned with the adverse economic and eventual social and political short-run consequences of this action. On the other hand, policy-makers are often too sensitive to the negative effects.

A well-designed package of measures that includes price increases and aggregate demand policies can alleviate if not avoid adverse consequences while putting the economy on a sounder growth track. Presumably, there are a number of policy packages that can have the desired effects; this paper presents one workable option. The previous sections analyzed how each policy in the package individually would affect the economy in the short run. The section considers the combined effects of implementing an entire package, assuming a 1 percent change under each policy instrument. In principle, the scope of the change through each policy can, however, be set wherever desired.

Column 1 of Table 3.7 presents the percentage change in selected macroeconomic indicators in response to a 1 percent increase in all public sector prices. The predominant effect -- a contraction of the economy -- prevails. The private sector as a whole and exports decline through the mechanisms described previously. Those two effects decrease imports and real private investment. Relative price changes benefit public real consumption

Table 3.7 - Combined macroeconomic effects of the complete package

	Public sector price increases	Aggregate demand policies	Complete package
A. Selected constant price aggregates			
GDP	-.262	.140	-.122
Private consumption	-.378	.179	-.199
Public consumption	.026	-.027	-.001
Public investment	.136	-.0334	.102
Private investment	-.055	.833	.778
Total investment	.053	.245	.298
Imports GNFS	-.026	.300	.274
Exports GNFS	.237	.291	.054
B. Other macroeconomic variables			
Current account deficit	.035	1.23	1.27
Parallel exchange rate	.523	-.006	.517
Urban wage	.000	.000	.000
Rural wage	.195	.630	.435
Urban CPI	.068	.365	.297
Rural CPI	.206	.444	.238
Employment	.018	.090	.072
C. Selected public finance indicators			
Government revenue	.100	.159	.259
Indirect taxes	.194	.290	.484
Direct taxes	-.057	.215	.158
Subsidies	-.069	.236	.167
Fiscal	-.333	.433	.100

a/ the figures indicate the percentage change in the indicators for a 1 percent change in the indicators. A is derived from a perturbation of the solution in 1983/84, while B and C are the outcome of perturbations of the solution in 1982/83.

and investment, which expand. The fiscal gap improves.

To summarize, raising public sector prices:

(1) results in a larger supply of public goods that displace private goods, forcing their prices down;

(2) has, in varying degrees, an adverse effect on exports that reduces the availability of foreign exchange in the commercial bank and central bank pools;

(3) cuts the amounts of imports because of tighter rationing attributable to the lower availability of foreign exchange and the depreciation of the parallel market exchange rate;

(4) squeezes the private sector between declining output prices and increasing costs of imported intermediates and public sector goods;

(5) creates an improvement in public sector finances through higher profitability of public sector firms and the windfall of import premia accruing to them; and

(6) reduces most distortions in the economy except import premia.

The response of the economy to the set of aggregate demand policies is shown in column 2 of Tables 3.7 and 3.8. Those policies cause an expansion, with an increase in GDP and all its components, except real public consumption and investment at constant market prices. The fiscal gap worsens in spite of greater public sector revenues. Any additional aggregate demand can only be satisfied by the private sector, imports or the demand-driven public sector. Any expansion in the output of the latter also feeds the demand for private sector goods and imports. As a consequence, private

Table 3.8 - Combined effects of the complete package
on levels of distortions a/

	Public sector price increases	Aggregate demand policies	Complete package
Shares of rents in GDP			
Subsidized consumer goods			
Rents accruing to urban households	-.917	1.89	.973
Rents accruing to rural households	-2.64	2.12	-.520
Total rents of households	-2.15	2.05	-.100
Public sector employment	-4.74	-2.02	-6.76
Public sector outputs markets	-4.74	-2.02	-6.76
Supply-driven sectors	-7.52	3.53	-3.99
Demand-driven sectors	-5.89	-4.38	-1.67
Imports premia	2.71	-4.38	-1.67

a/ the figures indicate the percentage change in the indicators for 1 percent change in the prices of the sectors shown in the columns. The elasticities are obtained from a perturbation of the 1982/83 solution.

sector prices and outputs expand, pulling private consumption, investment and imports with them. The combination of the switching policy and the resultant devaluation of the commercial bank exchange rate and of the absorption policy of investment expansion leaves the parallel market exchange rate almost unchanged. The higher public sector investment budget, lower import premia and higher subsidy payments cause the fiscal gap to widen.

The outcome in terms of the distortions is mixed. The rents accruing to households and the distortions in the output markets of the public sectors increase. Distortions caused by the fixed public sector wage bill and the rationing of imports are reduced.

The last columns of Tables 3.7 and 3.8 give the effects of the combination of price increases and aggregate demand policies. The latter policies reduce the adverse effects of the increases in public sector prices. The decline in real private consumption is less pronounced, while real public consumption is left almost unchanged. All other components of GDP, especially investment, increase. Combining the policies results in a lower decline in GDP without exacerbating the fiscal deficit, which increases only slightly. Finally, most distortions are alleviated.

The following can be concluded:

(1) It is important to recognize the limited response of the public sector and the more flexible one of the private sector.

(2) The response of the economy depends on the initial situation in the market in which the price increases takes place; there are supply-driven sectors in which demand is rationed, and demand-driven sectors in which adjustment occurs more in the operating surplus of the producers.

(3) The outcome also differs according to the nature of the commodity produced by the sector whose price is raised; results differ depending on the tradable or non-tradable nature of the commodity.

(4) It is important to identify the foreign exchange pool through which exports pass; if the exchange rate is flexible, the eventual adverse effect on exports is less pronounced than if it is fixed.

(5) The rationing of imports in the commercial bank pool is significant to the response of the rest of the economy and to the determination of the public sector gap.

(6) For each individual policy, the effects on the distortions are mixed, although the package as a whole reduces the distortions in almost every area.

(7) It is important to consider price increases as part of a package of measures that include aggregate demand policies, and not as individual price increases.

Chapter IV

THE MEDIUM TERM PERSPECTIVE OF A PUBLIC SECTOR REFORM PROGRAM

The previous chapter analyzed the short-term--within one year--response of the economy to individual and a combination of various policy measures designed to improve the efficiency of the public sector in Egypt. The main conclusion is that individual policy measures involve trade-offs among different indicators--some improve, others deteriorate. A combined policy package helps improve the trade-offs considerably. The short-run adjustment costs are still there, but they are softened significantly.

This chapter looks at the medium-term (over a period of 10 years) response of the economy to a public sector reform program. In this medium-term capacities expand, and wages and the supply of labor respond to the changes in the levels of activities and allocation signals that emerge from the the policy package. The reform is designed to be implemented over 10 years. During the first five years, the policy regime of 1983 is maintained, but with controlled adjustment of prices, as well as implementation of macroeconomic aggregate demand policies. The purpose is gradually to increase public sector prices in order to reduce distortions. Beyond the fifth year, assuming that the economy by then has acquired flexibility and the agents are now more familiar with price changes, public sector prices are freed to respond either to excess demand or, as with oil prices, to follow world prices. Simultaneous to the freeing of public sector prices, transactions with the rest of the world are consolidated into two pool, instead of three; the commercial bank and parallel markets for foreign exchange would be consolidated into one market, with the exchange rate free to adjust. The central bank is allowed to buy and sell foreign exchange from the free market. One distortion that remains is the difference in exchange rates governing transactions in the two pools, i.e. the free market and the central bank pool.

IV.1 PHASE I: MEDIUM TERM RESPONSE TO CONTROLLED ADJUSTMENTS

The policy package underlying the controlled adjustment phase consists of the following:

(1) Increases in the prices of all public sector production activities;

(2) Additional foreign borrowing to ease the adjustment process;

(3) Depreciation of the commercial bank pool exchange rate;

(4) Increased nominal investment in state enterprises to offset the deflationary impact of raising public sector prices (demand management);

(5) Lower borrowing of the public sector from the private sector in view of the additional public savings generated by the pricing reform (demand management).

The assumptions in this experiment are compared with those in reference path in Table 4.1.

Table 4.1: Key Assumptions Compared with the Reference Path
(rate of growth unless otherwise specified)

	Controlled Adjustments				Reference Path			
	83/84	84/85	85/86	86/87	87/88	83/84	84/85	85/86	86/87	87/88
Fiscal policy									
Share of private savings mobilized by the public sector	0.47	0.40	0.30	0.20	0.10	0.47	0.46	0.45	0.45	0.45
Nominal investment in state enterprises	0.3	0.28	0.28	0.25	0.25	0.23	0.25	0.23	0.21	0.20
Pricing policy									
Domestic price of petroleum products	0.10	0.165	0.165	0.165	0.60	0.10	0.11	0.12	0.12	0.12
Public food processing	0.075	0.165	0.165	0.165	0.165	0.075	0.075	0.06	0.06	0.06
Public textiles	0.090	0.165	0.165	0.165	0.165	0.09	0.075	0.06	0.06	0.06
Public other industries	0.090	0.165	0.165	0.165	0.165	0.09	0.075	0.06	0.06	0.06
Public agriculture	0.070	0.165	0.165	0.165	0.165	0.07	0.07	0.07	0.07	0.07
Public services	0.10	0.165	0.165	0.165	0.165	0.10	0.09	0.08	0.08	0.08
Electricity	0.05	0.165	0.165	0.165	0.165	0.05	0.05	0.05	0.05	0.05
Transport and communications	0.10	0.190	0.190	0.190	0.190	0.10	0.09	0.09	0.085	0.085
Construction	0.08	0.165	0.165	0.165	0.165	0.08	0.07	0.07	0.07	0.07
Exchange rate policy									
Depreciation of the commercial bank exchange rate	0.05	0.070	0.070	0.070	0.070	0.05	0.045	0.040	0.040	0.040
Foreign borrowing	0.32	0.42	0.24	0.21	0.17	0.170	0.170	0.160	0.15	0.13

(0963J-12)

1.a RESULTS OF THE EXPERIMENT

The growth scenario during the first five years of the reform program is presented in Table 4.2. The overall impact of the policy package on growth is deflationary. The decline in growth is induced by the increase in public sector prices which reduce both overall domestic demand and demand for exports. Policies to combat this deflationary impact, e.g., increased investment by public companies, reduced public borrowing and depreciation of the commercial bank exchange rate, fail to offset the deflationary effect fully. Reduced public borrowing enables higher private investment and consumption and thereby offsets the adverse impact of the public sector price increases on private demand. Nevertheless, despite the higher nominal investment by public companies, real public investment grows more slowly than in the reference case because of higher investment prices induced by the exchange rate depreciation and public sector price increases (see Table 4.2). Similarly, public consumption grows much more slowly as compared with the reference case. On the other hand, the depreciation of the commercial bank exchange rate fails to offset fully the disincentive effect that increased public sector prices have on exports. Although oil exports grow faster because of lower domestic demand, which releases additional exportable surplus, non-oil commodity exports still slow down. The reason is mainly the price increases in the public sector, which lower the export demand for price-sensitive commodities, but it is also the result of slower depreciation in the free market exchange rate which is a key determinant of private exports.

Table 4.2: Growth Scenario with Controlled Adjustments Compared with Reference Path in the Medium Term
(average annual real growth rates

	Reference path 1982/83 - 1987/88	Controlled adjustment 1982/83 - 1987/88
Household consumption	5.8	6.1
Government consumption	3.4	1.8
Total consumption	5.2	5.1
Public sector investment	3.9	2.6
Private investment	4.4	6.5
Total investment	4.1	4.4
Exports (GNFS)	6.5	5.0
Imports (GNFS)	3.4	5.1
GDP (at mp)	5.8	4.9
Employment	2.3	2.3

Table 4.3: Key Price Developments with Controlled Adjustments Compared with the Reference Path in the Medium Term
(Percentage Growth Rates)

	Reference path 1982/83 - 1987/88	Controlled adjustments 1982/83 - 1987/88
Wages (nominal)		
Public sector (Urban)	13.6	13.6
Private sector (Urban)	15.4	14.1
Public sector (Rural)	13.6	13.6
Private sector (Rural)	19.0	15.8
Prices		
Consumer (Urban)	13.6	12.2
Consumer (Rural)	15.0	12.8
Investment	16.7	18.6
Exchange Rates		
Commercial	4.3	6.6
Free market	9.4	5.7

One important factor contributing to the slower depreciation of the free market exchange rate is the increase in official foreign borrowing which augments the overall supply of foreign exchange in the economy despite a slowdown in export growth. On the whole, despite a large private demand, a slowdown in public sector demand and exports demand deflates the economy.

The savings-investment balance and fiscal indicators in the controlled adjustment path show some improvement over the reference path (see Tables 4.4 and 4.5). The overall fiscal deficit is lower, and total public savings are much higher. Private saving declines as a result of a reduction in public borrowing from the private sector. Part of the resources thus released feeds into consumption and thereby lowers overall private savings. The increase in public savings, however, more than offsets the decline in private savings to yield higher overall national savings. The savings of both public companies and government improve. The main factor contributing to higher savings by public companies is price increases, which more than offset the adverse consequences of a fall in import premia on the income and savings of public companies. Government savings improve because of lower growth in government consumption and an increase in tax revenues resulting from higher private income and consumption. Together they cause a significant decline in the fiscal deficit, despite higher nominal investment and higher subsidies. An important result is that, in spite of the significant cost-push pressures, overall inflationary pressures are reduced because of a slowdown in income growth and a lower fiscal deficit.

A major reason the Egyptian authorities have been reluctant to adjust public sector prices is a fear that this action would generate higher inflation. The results indicate that the general equilibrium effects of

Table 4.4: Savings-Investment Balance with Controlled Adjustments Compared with the Reference Path in the Medium Term
(Share of GDP at market prices)

	Reference path 1982/83	Reference path 1987/88	Controlled adjustment 1987/88
Total investment	27.5	28.4	34.8
Household savings	12.1	12.8	11.8
Private companies savings	3.3	3.6	3.3
- Total private savings	15.4	16.4	15.1
Public companies savings	6.7	6.6	7.2
Government savings	(-5.7)	-4.7	0.0
Social security savings	4.7	4.7	4.6
- Total public savings	5.7	6.8	11.8
- Total foreign savings	5.9	4.9	4.5

Table 4.5: Selected Fiscal Indicators with Controlled Adjustment Compared with the Reference Path in the Medium Term
(share of GDP at market prices)

	Reference path 1982/83	Reference path 1987/88	Controlled adjustment 1987/88
Total current expenditure	21.3	19.0	18.6
Subsidies	7.6	6.0	6.6
Total tax receipts	21.7	19.4	22.3
Public economic sector surplus	9.6	7.0	8.8
Total public investment	18.8	19.2	21.0
Overall fiscal resource gap	17.3	16.8	13.4
Foreign borrowings	5.5	4.6	7.2
Social security savings	4.7	4.7	4.6
Domestic private financing	7.1	7.5	1.5

public sector price increases are <u>deflationary</u>, and not <u>inflationary</u> 11/.

Another improvement in the controlled adjustment scenario is that the impact of the distortions is lower. Import premia, rents to consumers of subsidized goods and rents in the public sector output markets are all below levels obtaining in the reference path (see Table 4.6). The import premia fall because of the higher total supply of foreign exchange accruing from the additional foreign borrowing which more than offsets the impact of a slowdown in import earnings. The share of rent accruing to households declines because increased private investment and lower private sector real wages expand output in the private sector, which reduces prices in the private sector and thus lowers the gap between procurement and subsidized prices. On the other hand, the increase in public sector prices reduces the gap between virtual and market prices and thereby helps reduce distortions in the goods market. Although demand management policies tend to weaken the improvement in distortions in the goods market, the final outcome is a moderate improvement in rent.

The distortion in the public sector employment market is, however, aggravated somewhat. The main reason is the rigidity of employment in that sector. Although output in the public sector contracts, employment remains unchanged because of policy. The gap between marginal productivity and the regulated wage increases.

11/ The model does not include an explicit monetary sector. Thus it is not strictly valid to talk about inflation or deflation. However, the relevance of the result that a reform of public sector pricing policy is <u>deflationary</u> rather than <u>inflationary</u> is likely to remain valid even when a monetary sector is introduced explicitly. A reduction in private income will lower the demand for goods and services in the economy and thus provide one offset to the cost-push pressures. More important, a lower public sector budget deficit means a lower money supply which will also contribute to the deflationary impact of a public sector pricing reform.

To summarize, during the controlled adjustment phase there will be an overall improvement in the macroeconomic balance of the economy, but the adjustment costs in terms of lower economic growth will prevail in the medium term. There will be some improvement in the efficiency of resource utilization, although, again, the efficiency gains will be somewhat limited in the medium term.

Table 4.6: Indicators of distortions with controlled adjustment Compared with the reference path in the medium term (share of GDP)

	Reference path 1982/83	Reference path 1987/88	Controlled adjustment 1987/88
Share of rents in GDP			
Subsidized consumer goods:			
Rents to urban household	0.5	0.7	0.6
Rents to rural household	1.2	1.8	1.5
Total rents	1.7	2.5	2.1
Public sector employment	1.1	2.0	2.4
Import premia	4.2	9.3	2.9
Public sector output markets:			
(i) Supply driven sectors	14.1	19.8	17.1
(ii) Demand driven sectors	1.3	1.2	0.2

IV. 2 Phase II: Macroeconomic Trends in a Flexible Policy Environment

The main assumptions underlying the reform program in the second phase are as follows:

(1) Government intervention in the markets for public sector output is eliminated. Thus, all commodity prices are now market-determined;

(2) The exchange system is liberalized;, the commercial bank and free markets are unified, and the exchange rate is left free to be determined by demand and supply.

(3) Government controls on public sector employment are eliminated so that public companies are free to hire and fire and wages are determined by the market.

To capture fully the implications of liberalization, it is assumed that in the second phase (i) public enterprise investment is at the same level as in the corresponding period of the reference path and (ii) foreign borrowings are also at the same level as in the reference path.

2.a. Liberalization Experiment Results

There is substantial improvement in the overall growth prospects (see Table 4.7). Average GDP and consumption grow by 8.4 and 7.4 percent respectively, compared with only 4.8 and 4.4 percent in the reference path. The two main vehicles of this growth are the surge in exports and the improved efficiency of resource use. The liberalization measures also eliminate many of the constraints on exports; thus, the unified exchange rate depreciates faster than the differential between world and domestic inflation, a situation that improves the export incentives in the economy. At the same time, the improved efficiency of production helps lower the price increases throughout the economy. For both reasons, that is, the exchange rate depreciation and the lower price increases, exports expand dramatically--by 15.0 percent compared with only 5.1 percent in the reference path. Investment growth is virtually unchanged: private investment increases

Table 4.7: Growth scenario under a more flexible policy Regime compared with the reference path
(Annual average growth rate)

	Reference path (1987/88 - 1992/93)	Combined adjustment (1987/88 - 1992/93)
Household consumption	4.7	7.1
Government consumption	3.6	8.3
Total consumption	4.4	7.4
Public sector investment	3.0	1.8
Private sector investment	2.9	4.9
Total investment	3.0	3.1
Exports (GNFS)	5.1	15.0
Imports (GNFS)	2.7	4.0
GDP (at mp)	4.8	8.4
Employment	2.4	2.0

faster, public investment advances more slowly. Further, employment growth is also lower. The achievement of higher economic growth with a slower increase in both investment and employment illustrates the impact of the efficiency gains resulting from liberalization.

The dramatic improvement in growth prospects even with a slower investment growth highlights the importance of eliminating the sources of distortion in the Egyptian economy. In an environment riddled with major policy induced constraints, the actual productivity of key resources, foreign exchange, labor and capital, are much below their true levels. Consequently, when the distortions are removed, the realized productivity gains are substantial. The factors of production are now mobile and respond to market signals which now better reflect true economic values. Resources move from low productivity activities to high productivity activities enabling both higher output and lower costs. This is the key message of the liberalization experiment.

The savings-investment balance and the fiscal situation also improve (see Table 4.8 and 4.9). As noted, private savings decline because of lower public sector borrowing, but public savings improve dramatically because of higher tax revenue induced by private sector income growth and a larger fiscal role of the public economic sector, which is no longer financially

Table 4.8: Savings-Investment balance under a flexible policy Regime compared with the reference path
(Share of GDP at market prices)

	Reference path 1991/92	Combined adjustment 1991/92
Total investment	28.5	32.9
Household savings	12.6	11.5
Private companies savings	3.5	3.0
Total private savings	16.1	14.5
Public companies savings	7.0	7.7
Government savings	-3.6	-0.3
Social security savings	4.7	4.6
Total public savings	7.9	12.0
Total foreign savings	4.5	6.4

Table 4.9: **Selected fiscal indicators under a flexible policy Regime compared with the reference path**
(share of GDP at market prices)

	Reference path 1991/92	Combined adjustment 1991/92
Total current expenditure	18.6	22.1
Subsidies	5.3	6.2
Total tax receipts	19.1	24.4
Public economic sector surplus	5.6	8.9
Total public investment	19.4	19.6
Overall fiscal resource gap	16.2	12.1
Foreign borrowings	4.1	-
Social security savings	4.7	4.6
Domestic private financing	7.4	-

Table 4.10: **Key price developments under a flexible policy Regime compared with the reference path**
(annual percentage change)

	Reference path (1986/87-1991/92)	Controlled adjustments (1986/87-1991/92)
Wages (Nominal)		
Public sector (Urban)	10.7	9.8
Private sector (Urban)	12.2	6.4
Public sector (Rural)	10.7	9.8
Private sector (Rural)	15.3	9.6
Prices		
Consumer (Urban)	10.7	3.9
Consumer (Rural)	11.6	4.2
Investment	13.5	10.0
Exchange Rate		
Commercial bank	4.0	-
Parallel market	7.7	7.7 a/

a/ Unified exchange rate.

constrained. Public consumption and subsidies grow faster, but the increased public sector savings induce a substantial improvement in the overall fiscal balance. Inflationary pressures are now much reduced (see Table 4.10). The large gains in efficiency cause a significant expansion in output, which lowers prices in the economy, another striking result. High growth and low inflation are both achievable targets, and, moreover, price flexibility and low inflation are not necessarily contradictory objectives.

As expected, the bulk of the rents in the economy are eliminated (see Table 4.11). By freeing the economy, the policy program naturally has removed the key sources of distortions. Households do, however, continue to derive some rent from their access to rationed goods, but the distortionary impact is small.

Table 4.11: Indicators of distortions under a flexible policy Regime compared with the reference path
(share of GDP at market prices)

	Reference path 1991/92	Combined adjustment 1991/92
Share of Rents in GDP		
Subsidized Consumer Goods:		
Rents to urban household	0.67	0.13
Rents to rural household	1.78	0.21
Total rents	2.45	0.34
Public sector employment	1.90	-
Import premia	11.0	-
Public sector output market		
Supply-driven sectors	19.7	-
Demand-driven sectors	0.8	-

To summarize, with bold initiatives involving a reform of the pricing policy and complementary demand measures, Egypt can achieve high rates of growth and improve the overall macroeconomic balance. The objective of the policy reform should be to correct the distortions that have accumulated over time because of government interventions. In the short to medium term there will be some adjustment costs in terms of lower economic growth. However, the gains in efficiency in the long run will create substantial improvement in Egypt's growth prospects and strengthen its macroeconomic balance as a whole.

Chapter V

SUMMARY

Egypt's long-term growth prospects depend crucially on reform of the numerous interventions that constrain the performance of public enterprises. The three major sources of distortions are the prevalence of multiple exchange rates, price controls on public sector output and interventions in public sector employment decisions. These distortions have been creating severe inefficiencies in the utilization of resources and have limited the ability of the public sector to generate additional resources. The simulation results based on a detailed macroeconomic framework, MISR2, which is designed to capture the impact of the major distortions, show that adjusting public sector prices alone will not yield unambiguous results: some indicators improve, others deteriorate. On the other hand, a flexible policy environment in which most of the sources of distortions have been removed will lead Egypt to a high growth path and improve the macroeconomic balance.

A flexible environment that allows mobility of factors and yields price signals which better reflect true economic values, will raise the productivity of resources and thereby improve overall growth prospects. A more flexible environment also improves the macroeconomic balance. The flexibility of the exchange rate, all other prices and factors of production, significantly improve incentives for exports, while imports fall responding to the more realistic cost of foreign exchange. Both result in an improvement in the balance of payments. On the other hand, the combined effects of larger government revenue, due to higher economic growth, and lower subsidy expenditure, because of the elimination of price controls, improve the fiscal situation.

Another important conclusion is that there is no logical contradiction between price flexibility and dampening of inflationary pressures. Price flexibility will contribute to both greater efficiency in resource use and reduce the budget deficit by raising public savings. Nominal aggregate demand will fall, whereas real aggregate supply will rise, causing a decline in inflationary pressures.

It should be reiterated that although price flexibility is an essential ingredient in a policy reform program, it has to be combined with flexible exchange rate and public sector employment policies in order to achieve the desired improvement in growth and macroeconomic balance.

APPENDICES

APPENDIX I
The Within-Period Module of the MISR2 Reference Model

Page No.

INTRODUCTION 99

CHAPTER 1: The Economics of the MISR2 Reference Model 99

    1.   The Goods Markets 100

        1.1 From Original Suppliers to Traders 100
            1.1.1 From Domestic Suppliers to Traders 101
            1.1.2 From Foreign Suppliers to Traders 107

        1.2 From Traders to Final Users 110

    2.   Factor Markets 110

    3.   Foreign Exchange and Trade 112

    4.   Macroeconomic Equilibrium 113

CHAPTER 2: The Formulation of the MISR2 Reference Model 115

A. The TV Approach 115

    1.   An Overview of TV 115

B. The MISR2 SAM 116

    1.   An Introduction to the SAM Framework 116

    2.   The Aggregate SAM for Egypt 118

    3.   An Intermediate Disaggregation of the SAM Accounts 120

        3.1 An Intermediate Disaggregation of Activities
            and Commodities 121

        3.2 An Intermediate Disaggregation of the Rest of
            the World Accounts 125

4. Disaggregated Individual Accounts of the MISR2
   Model and Its SAM 128

   4.1 Factors of Production 128
       4.1.1 Labor 128
       4.1.2 Capital and Land 131

   4.2 Composite Inputs 134

   4.3 Current Institutions Accounts 139
       4.3.1 Household Expenditures 139
       4.3.2 Companies Accounts 139
       4.3.3 Government, Social Security, and the
             Tax Accounts 142
       4.3.4 The Rest of the World Accounts 142

   4.4 Capital Account 145

   4.5 Activities 145
       4.5.1 Public Activities 145
       4.5.2 Private Activities 148

   4.6 Commodities 148
       4.6.1 Government Trade Commodities
             (Domestic, Imports, and Exports) 148
       4.6.2 Non-government Commodities
             (Domestic and Exports) 151
       4.6.3 Non-government Imports 154

C. The MISR2 TV 154

   1. The TV Specifications used in MISR2 156

   2. The Within-Period Module of the Reference Model in TV 158

CHAPTER 3: The Alternative MISR2 Model: Changes in Policy Regime 179

   1. Introduction 179

   2. Liberalizing Public Sector Output Prices 179

   3. Public Sector Employment 182

   3. Reforming the Exchange Rate and Trade Regime 182

APPENDIX I, CHAPTER 1
LIST OF TABLES

Table No.		Page No.
3.1.A	From Original Suppliers to Traders	102
3.1.B	From Original Suppliers to Traders	103
3.2.A	From Traders to Final Users	111
3.2.B	From Traders to Final Users	112

APPENDIX I, CHAPTER 2
LIST OF TABLES

Table No. | Page No.

A.1.a   Aggregate SAM for Egypt -- SAM 1, 1979 119
A.1.b   Intermediate SAM for Egypt--Disaggregation of
          Activites and Commodities 122
A.1.c   Intermediate SAM for Egypt--Disaggregation of
          the Rest of the World 126

FACTORS OF PRODUCTION

A.1.1   Urban Labor 129
A.1.2   Rural Labor 130
A.1.3   Total Labor 132
A.1.4   Capital and Land 133

COMPOSITE INPUTS

A.1.5   Composite Domestic Intermediate Inputs: Public Activities 135
A.1.6   Composite Domestic Intermediate Inputs: Private Activities 136
A.1.7   Composite Imported Intermediate Inputs: Public Activities 137
A.1.8   Composite Imported Intermediate Inputs: Private Activities 138

CURRENT INSTITUTIONS ACCOUNTS

A.1.9   Household Accounts 140
A.1.10 Companies Accounts 141
A.1.11 Government, Social Security, and the Tax Accounts 143
A.1.12 Rest of the World Accounts 144

CAPITAL ACCOUNT

A.1.13 Capital Accounts 146

ACTIVITIES

A.1.14 Public Activities 147
A.1.15 Private Activities 149

COMMODITIES

A.1.16 Government Trade Commodities (Domestic, Imports, & Exports) 150
A.1.17 Non-Government Commodities (Domestic and Exports) 152
A.1.18 Non-Government Composite Commodities (Domestic and Exports) 153
A.1.19 Non-Government Imports 155

APPENDIX I, CHAPTER 2
LIST OF TABLES [1]

Table No.		Page No.

FACTORS OF PRODUCTION
- A.2.1   Urban Labor 160
- A.2.2   Rural Labor 161
- A.2.3   Total Labor 162
- A.2.4   Capital and Land 163

COMPOSITE INPUTS
- A.2.5   Composite Domestic Intermediate Inputs: Public Activities 164
- A.2.6   Composite Domestic Intermediate Inputs: Private Activities 165
- A.2.7   Composite Imported Intermediate Inputs: Public Activities 166
- A.2.8   Composite Imported Intermediate Inputs: Private Activities 167

CURRENT INSTITUTIONS ACCOUNTS
- A.2.9   Household Accounts 168
- A.2.10 Companies Accounts 169
- A.2.11 Government, Social Security, and the Tax Accounts 170
- A.2.12 Rest of the World Accounts 171

CAPITAL ACCOUNT
- A.2.13 Capital Accounts 172

ACTIVITIES
- A.2.14 Public Activities 173
- A.2.15 Private Activities 174

COMMODITIES
- A.2.16 Government Trade Commodities (Domestic, Imports, & Exports) 175
- A.2.17 Non-Government Commodities (Domestic and Exports) 176
- A.2.18 Non-Government Composite Commodities (Domestic and Exports) 177
- A.2.19 Non-Government Imports 178

---

[1] Tables A.2.1 to A.2.19 correspond to tables A.1.1 to A.1.19. The difference between those two sets of tables is that the former represent the accounting structure of MISR2 models, while the latter present TV specifications and account types.

APPENDIX I, CHAPTER 3

LIST OF TABLES 1/

Table No.		Page No.
A.3.14	Rest of the World Accounts	180
A.3.17	Public Activities	181
A.3.18	Non-Government Commodities (Domestic and Exports)	183
A.3.3	Non-Government Composite Commodities (Domestic and Exports)	184
A.3.12	Total Labor	185
A.3.19	Non-Government Imports	186

1/ Tables A.3.3 to A.3.19 correspond to tables A.2.3 to A.2.19. The difference between these two sets of tables reflects changes in policy regimes.

INTRODUCTION

MISR2 is a class of economywide equilibrium models.[1] The models of the class have the same accounting framework, behaviors and technology. They mainly differ in their systems constraints or closures [2] which characterize the nature of the within-period equilibrium. These closures reflect essentially the policy regime and institutional arrangements. Thus two models in the MISR2 class will correspond to different policy regimes and institutional arrangements. A comparison of two paths derived from two models in the class will amount to compare the outcomes of alternative policy regimes.

One model within the MISR2 class has been identified to reflect the policy regime of early 1983. This is the reference model from which the reference path is derived. In the following of this appendix chapter 1 provides a non-technical presentation of the reference MISR2 model corresponding to the policy regime prevalent around the middle of 1983. In Chapter 2, a full description of the within-period module is given. Because MISR2 is implemented along the Transactions-Value (TV) approach, the chapter begins with a brief overview of the approach. Then the Social Accounting Matrix (SAM) underlying the model is presented. Finally the TV specifications and the systems constraints of MISR2 are provided. In Chapter 3, modifications of the reference MISR2 model are presented. They correspond to a change in policy regime which would make public sector prices flexibles, market oriented and which would unify the commercial banks and "parallel" markets of foreign exchange.

CHAPTER 1 - The Economics of the Reference Model

A within-period general equilibrium is the outcome of two sets of factors: (i) the independent decisions of the agents intervening in the economy and (ii) rules which ensure the consistency of all such decisions. What characterizes alternative policy regimes is the set of rules which impose

---

[1] See Blitzer, Clark and Taylor (1975) and Dervis, de Melo and Robinson (1982). MISR2 is a class of economywide equilibrium models. All the models in the class have the same accounting framework, behaviors and technologies. The models in the class correspond to alternative policy regimes. Basically MISR2 distinguishes nine production activities and two households categories. The production activities are: Agriculture, Food-Processing, Textiles, Other-Industry, Electricity, Construction, Oil, Transportaton and Communication and Services. Households are separated into rural and urban.

[2] We use alternatively "systems constraints" or "closures" to indicate the rule imposed on each market in the economy telling how the market clears. Thus there are closures for factor markets, product markets, foreign exchange markets. Taylor (1978) uses closures to describe the macroeconomic constraints in the economy. Ginsburgh and Robinson (1982) describe the way various markets clear as system constraints. See also Robinson (1983).

the reconciliation of agents decisions. One can classify these rules into four interdependent categories according to the markets to which they pertain or to the price which governs transactions on those markets. Thus one can identify rules related to (i) goods [1] markets; (ii) factor markets; (iii) foreign exchange markets and (iv) macroeconomic equilibrium. The definition of the rules "closing" these interdependent markets, leads to a particular model in the MISR2 class. This model will correspond to a particular set of institutional arrangements and to a policy regime. In the following we review the rules corresponding to the reference model.[2,3]

1. The Goods Markets

There are two sets of goods markets. In the first set, goods are traded between "original suppliers" and "traders." In the second set the trade takes place between "traders" and "final users." The two sets of markets are considered successively in the following.

1.1 From Original Suppliers to Traders

Original suppliers are domestic and foreign producers. In the MISR2 class of models, in order to reflect conditions in the Egyptian economy, there are public and private domestic producers. Similarly, foreign suppliers are classified into three categories, depending on the foreign exchange pool used to bring the imports into the country. The three foreign exchange pools are: (i) the central bank, (ii) the commercial banks and (iii) the parallel market. Thus over-all there are two domestic, plus three foreign sources of supply of goods. They are represented in the rows of Table 3.1.

Traders are represented in the columns of Table 3.1. They undertake one of the three following activities: (i) domestic trade, exchanging domestically produced goods on domestic markets; (ii) export and (iii) import. Again to reflect conditions in Egypt in early 1983, each of these activities is separated into two, depending on the existence or absence of regulations governing the activity. There are thus six categories of traders. In the following we consider first trade between domestic suppliers and traders and then trade between foreign suppliers and traders.

---

[1] "Goods" is used here in a loose way to denote goods and non-factor services.

[2] For the inter-period linkages, see appendix 2.

[3] There is an expanding literature on "closing rules" analyzing the implications of alternative closures. Whereas this literature sheds light on some of the issues involved, it has a major drawback in that all discussions on closures depend in an essential way on the initial model framework the analyst starts with. This point appears clearly when one compares the discussion in Lysy (1983) and in Dewatripont and Michel (1983).

1.1.1 From Domestic Suppliers to Traders

Each row of Table 3.1[1/] corresponds to a market where trade takes place between an original supplier and traders. The first nine lines refer to goods and services produced by public sector activities. These markets are characterized by fixed prices: prices faced by public sector activities are administered and used as policy instruments. The question which then arises is that of the clearing of these markets.

For all markets, except those for electricity, oil, and services, the clearing rule is described in Figure 3.1. In panel A, the supply curve is at S, and the administered price at p, leading to a level of supply at q which needs to be allocated between traders. In panel B, the demand curve for all traders, except those concerned with domestic trade on unregulated markets, [2/] is at ED, leading to a level of their demand at q_e. Domestic traders on unregulated markets obtain $q - q_e$. In terms of goods produced by public sector activities, these traders are rationed. The rule described in Figure 3.1 corresponds to a situation where prices are fixed and supplies are limited; demand has to adjust. In this situation, an adjustment in the price p, say upwards, will have as first effects: (i) an expansion of supply; (ii) a dampening of export and regulated domestic demand, and (iii) a lower rationing of non-regulated domestic demand. Of course these responses will depend on the elasticities of the curves S and ED. The specification of Figure 3.1 is adopted here to reflect the view that there are limitations on supply in the markets concerned. For instance, the supplies of public means of transportation, publicly produced industrial and agricultural goods are considered to be limited.

For electricity and services produced by the public sector, the supply constraint is assumed not to be binding. The working of these two markets is explained in Figure 3.2. MC is the marginal cost curve; it is not the supply curve which is the flat line S at the price p. Thus supply is assumed to be perfectly elastic at the administered price p. This assumption implies that both electricity and services operate below full capacity. The assumption of increasing marginal cost is there to capture the fact that when production is expanding, less efficient idle inputs are brought into production. Assume now the demand curve to be at D_2, the firm will supply q_2 at the price p. The marginal cost will be at M_2 and the profit margin of the firm will be squeezed. If, on the contrary the demand curve is at D_1, the firm would be making a pure profit beyond the return to capital measured by its marginal productivity. An adjustment in p will affect, in the first round, on the one hand the levels of demand and production and on the other hand the profitability of the firm. However, the output response is not directly related to the price adjustment but indirectly through demand.

One last market to be considered is that of oil. There both the price and the output are exogenously fixed. Figure 3.3 explains the operation of the oil market. Output is given at $\bar{q}$ and the price p values output at the

---

[1/] Note that Table 3.1 consists of two parts: Table 3.1.A and Table 3.1.B.

[2/] This demand curve is for foreign and regulated demand by the government.

Table 3.1.A

From Original Suppliers to Traders

Table 3.1.B

From Original Suppliers to Traders

		Imports																	
		Unregulated Channels of Distribution									Regulated Channels of Distribution								
		Agriculture	Food processing	Textiles	Other Industry	Electricity	Construction	Oil	Transportation & Communication	Services	Agriculture	Food processing	Textiles	Other Industry	Electricity	Construction	Oil	Transportation & Communication	Services
		1	2	3	4	5	6	7	8	9	1	2	3	4	5	6	7	8	9
Rest of the World: Central Bank Pool of Foreign Exchange	Agriculture 19										X								
	Food processing 20											X							
	Textiles 21																		
	Other industry 22				X									X					
	Electricity 23																		
	Construction 24																		
	Oil 25							X											
	Transport. & communica. 26								X										
	Services 27									X									
Rest of the World: Commercial Banks Pool of Foreign Exchange	Agriculture 28	X																	
	Food processing 29		X																
	Textiles 30			X															
	Other industry 31				X														
	Electricity 32																		
	Construction 33																		
	Oil 34							X											
	Transport. & communica. 35								X										
	Services 36									X									
Rest of the World: Parallel Market Pool of Foreign Exchange	Agriculture 37	X																	
	Food processing 38		X																
	Textiles 39			X															
	Other Industry 40				X														
	Electricity 41																		
	Construction 42																		
	Oil 43							X											
	Transport. & communica. 44								X										
	Services 45									X									

Figure 3.1

Supply Driven Markets
Between Original Suppliers and Traders

Panel A

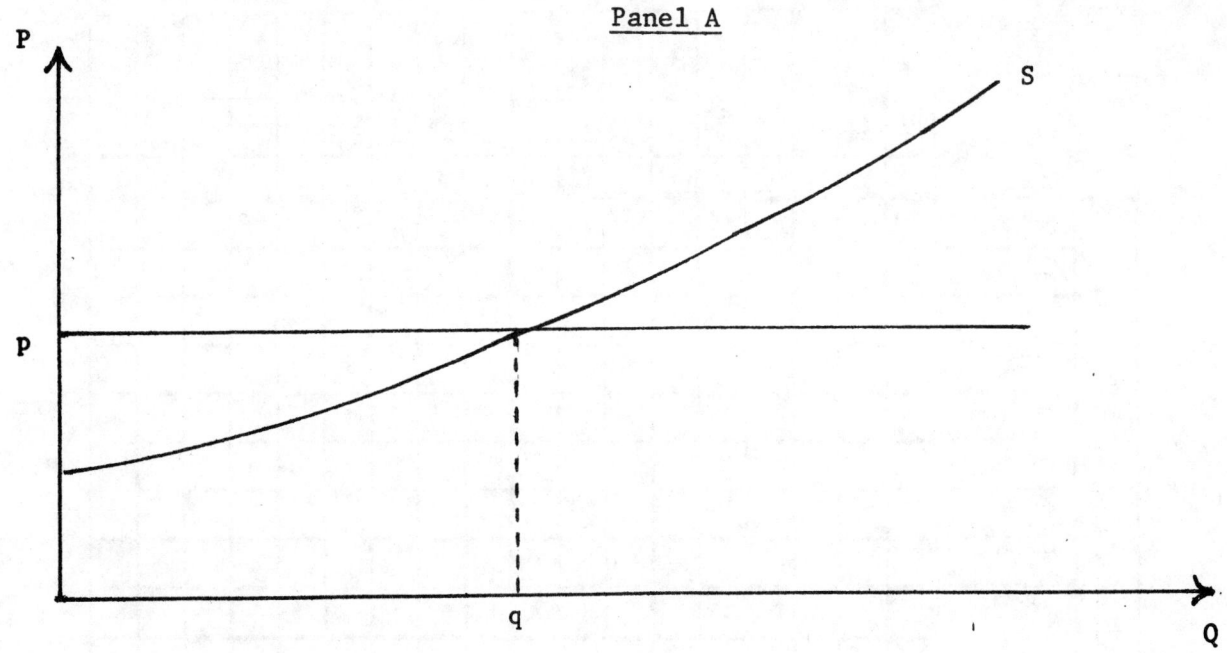

Panel B

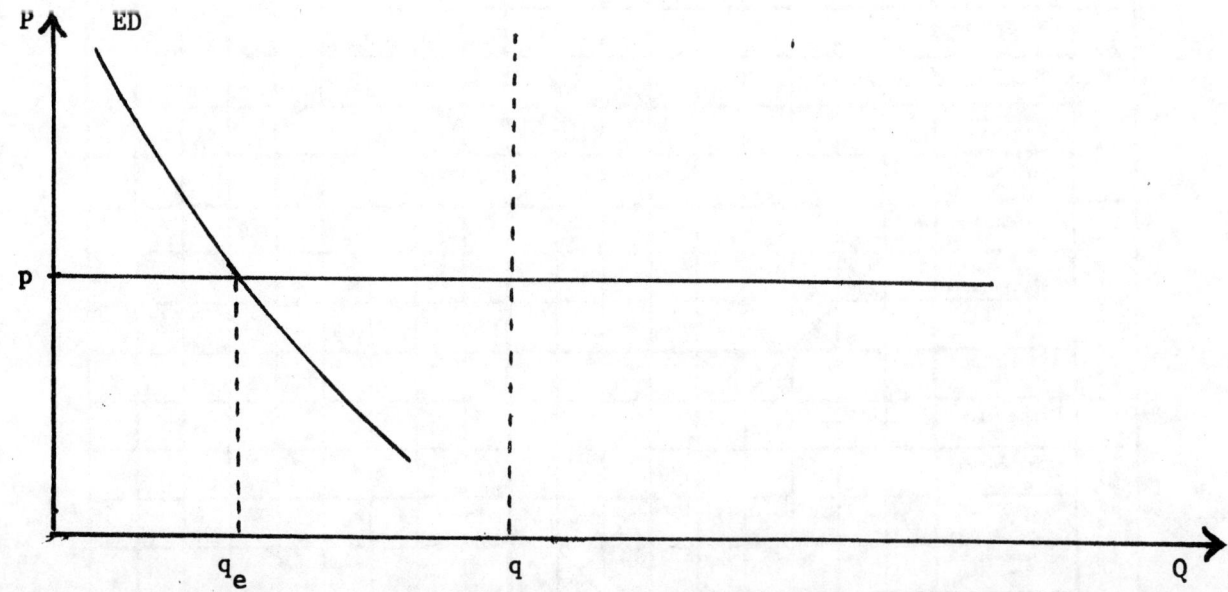

Figure 3.2

Demand Driven Markets: Electricity and Public Sector Services
Between Original Suppliers and Traders

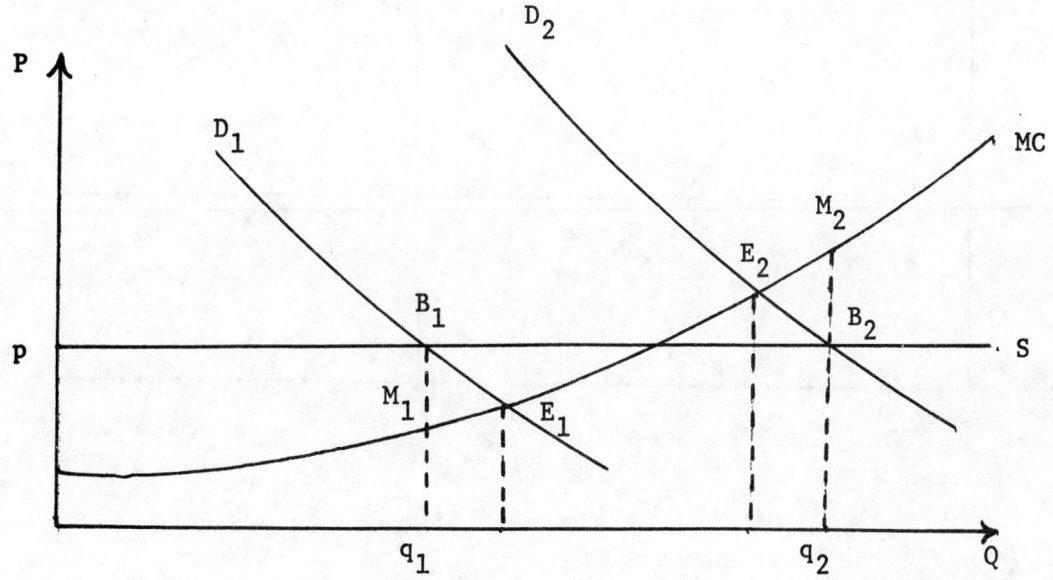

Figure 3.3

The Market of Oil

Between Original Suppliers and Traders

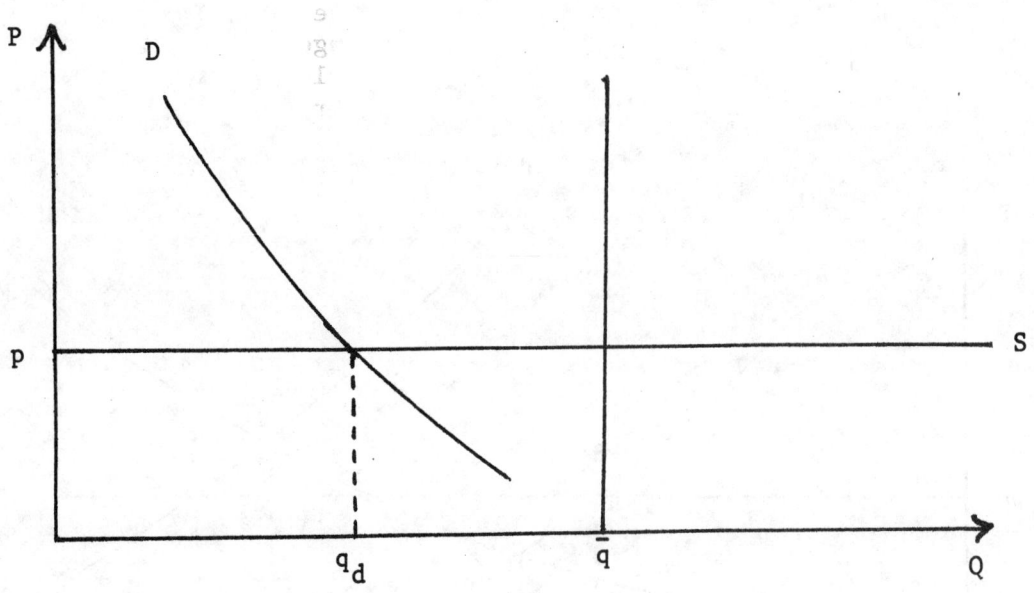

"well head." After allowing for taxes and subsidies, domestic demand is obtained from the intersection of the demand curve D and the supply curve S at q_d; $q - q_d$ is the quantity exported. Thus the oil market has a fixed domestic price, a fixed output and perfectly elastic export demand. Exports clear the market. An upward adjustment in the price p would tend to contract domestic demand and leave a larger exportable surplus.

In the foregoing we have dealt with the nine first lines of Table 3.1, corresponding to the nine markets where trade takes place between public sector production activities and traders. These markets are characterized by fixed prices and are either demand or supply driven. We now turn to the next nine lines in Table 3.1, corresponding to the markets where private producers sell their outputs. All private producers sell on competitive markets where the price clears supply and demand. This is not however completely true for private agriculture. Indeed government trading agencies will buy a certain amount of production at a regulated price. The level of activity of private agriculture will however be determined by the level of non-regulated demand: farmers will produce beyond what the government agencies will buy. The additional supply will be sold at a price which clears the market. Figure 3.4 explains the operation of the market of agricultural goods. Government agencies will buy q_g at the price p_g and $q_m - q_g$ will be sold on the free market at the price p_m.

1.1.2 From Foreign Suppliers to Traders

According to the exchange rate regime and the regulations which govern the transaction, imports can be classified into three categories: (i) those financed through the central bank foreign exchange pool, (ii) those financed through the commercial banks foreign exchange pool, and (iii) imports financed through the parallel market pool. 1/

Lines 18 to 26 of Table 3.1 refer to the markets of imports supplied through the central bank pool where the exchange rate is fixed. Along with the small country assumption the world supply, at the going exchange rate, is perfectly elastic. However traders cannot import any desired quantities through this pool. Regulated imports are fixed in quantities, corresponding to an inelastic demand by the importing agencies. The unregulated imports, except for oil, are rationed through a quota-type system. In Figure 3.5, panel A, the demand curve for imports is at D, the world supply at p_m and the quota at $\bar{q}$. There is thus a virtual price at p_v and an implicit rent corresponding to the shaded area. Panel B refers to imports of oil which are not submitted to a quota. Wherever the demand schedule D intersects the world supply curve at p_m, the implied imports will be allowed in. A devaluation of the exchange rate governing transactions on the central bank pool of foreign exchange will in a first round contract oil imports and the implicit rents on the other imports. 2/

1/ In early 1983 the Central Bank rate was at .7 LE per $, the commercial banks rate was at .85 LE while the parallel market was close to 1.2 LE.

2/ If the devaluation is too large it could transform the rent into a tax.

Figure 3.4

The Market for Privately Produced Agricultural Goods

Between Original Suppliers and Traders

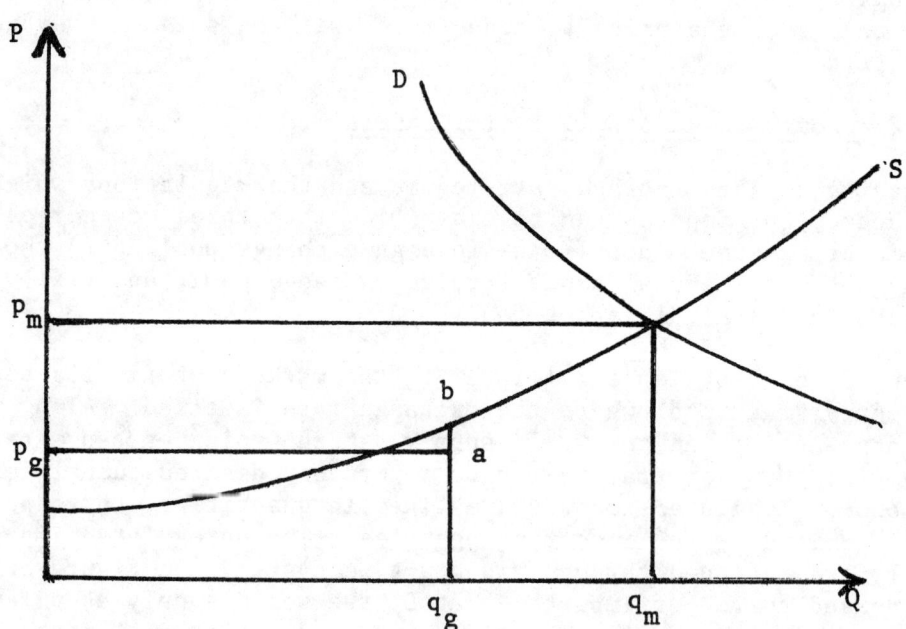

Figure 3.5

Panel A
Central Bank Pool
Non-Oil Imports

Panel B
Central Bank Pool
Oil Imports

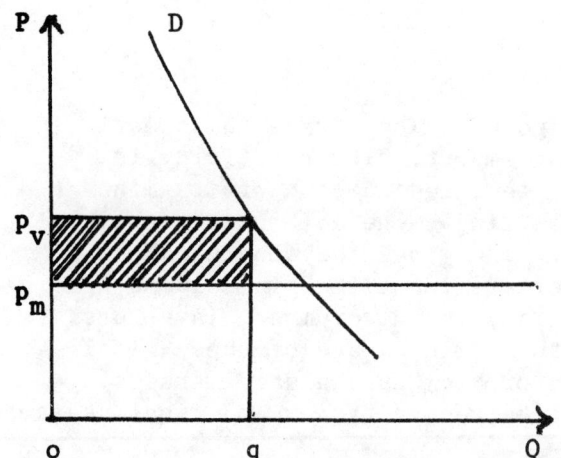
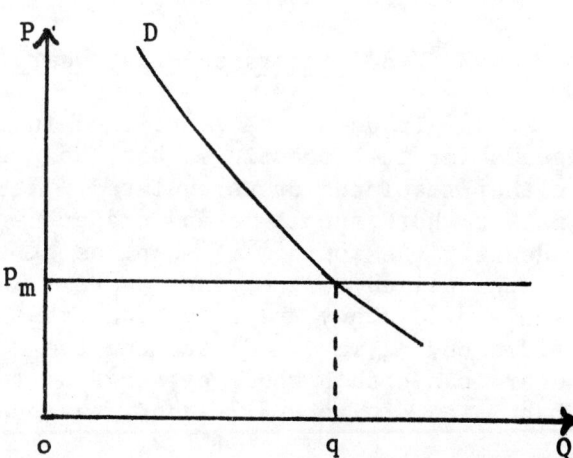

Figure 3.6

Commercial Banks Pool
All Imports

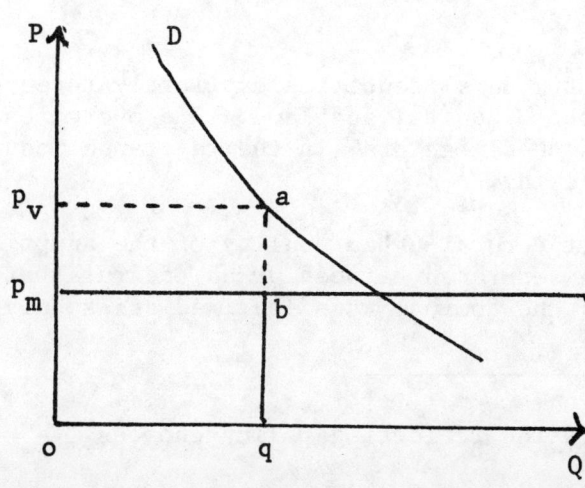

Lines 27 to 35 of Table 3.1 correspond to imports through the commercial banks pool where the exchange rate is also fixed. On this pool, imports are rationed but through a different mechanism than on the central bank pool. Here foreign exchange is allocated on different types of imports commodities.[1] Given the world price and the exchange rate, this allocation determines the quantities which can be imported. In Figure 3.6, q is the quantity which can be afforded given the world price and the exchange rate. A rent corresponding to $p_v ab p_m$ is implied. A devaluation of the commercial banks exchange rate will contract the rent.

On the "parallel market" pool (lines 36 to 44 of Table 3.1) the allocation of foreign exchange and the determination of the level imports are different. The exchange rate is free to adjust to clear the supplies and demand for foreign exchange.

1.2 From Traders to Final Users

Traders were identified in the foregoing. They trade in domestic goods for the domestic market, they export or import. Their activity is either regulated or unregulated. Table 3.1 presented in each of its lines a market where suppliers and traders are transacting. The columns of Table 3.1 identify the sources of supplies of traders. The goods they buy need now to be channelled to the final users. The latter are identified in columns of Table 3.2. They are rural and urban households, the government, investors, firms buying intermediates and the rest of the world. Each of these final users can obtain goods either from regulated or from unregulated markets. Each line of Table 3.2 identifies one market between a trader and final users.

The distinction between regulated and unregulated markets is the following. On the former prices are fixed and final users are rationed. On the latter prices are flexible and clear the markets. On the regulated markets both the ration and the price are policy variables. Exports through regulated channels of distribution are however not rationed. Indeed oil exports are a residual equal to what is left after domestic consumption is satisfied. Exports of transportation are in effect the receipts of the Suez Canal. They are exogenous in dollars depending on the state of world trade. For agricultural exports through regulated channels, the government trading agency procures fixed quantities from producers and tries to get the best price on the world market. The difference will find its way to the budget.

2. Factor Markets

The MISR2 class of models identifies explicitly three categories of factors of production: labor, capital and land. The operations of the markets of the services of these factors, in the reference model of the class, are considered in the following.

Corresponding to the rural-urban duality of the Egyptian economy, labor markets in MISR2 are separated between urban and rural areas. In urban areas, within each period, the nominal wage is fixed and supply is assumed

---

[1] This is a direct allocation reflecting policy choices.

Table 3.2.A
From Traders to Final Users

Traders		Households			Investment			Intermediates		R.O.W.		
	Final Users	Urban	Rural	Government	Government	Private	Public	Private	Public	Commercial Banks Foreign Exchange Pool	Central Banks Foreign Exchange Pool	Parallel Markets Foreign Exchange Pool
Domestic Markets — Unregulated Channels of Distribution	Agriculture	X	X		X			X	X			
	Food processing	X	X					X	X			
	Textiles	X	X					X	X			
	Other Industry	X	X					X	X			
	Electricity	X	X	X	X	X	X	X	X			
	Construction				X	X	X					
	Oil	X	X		X	X	X	X	X			
	Transportation & Communica.	X	X	X	X	X	X	X	X			
	Services	X	X	X	X	X	X	X	X			
Domestic Commodities — Regulated Channels of Distribution	Agriculture	X	X					X				
	Food processing	X	X									
	Textiles											
	Other Industry	X	X									
	Electricity											
	Construction											
	Oil											
	Transportation & Communica.											
	Services											
Exports — Unregulated Channels of Distribution	Agriculture									X		
	Food processing										X	
	Textiles										X	
	Other Industry										X	
	Electricity											
	Construction											
	Oil											
	Transportation & Communica.										X	
	Services									X		

Table 3.2.B
From Traders to Final Users

		Households			Investment			Intermediates		R.O.W.		
Traders / Final Users		Urban	Rural	Government	Government	Private	Public	Private	Public	Commercial Banks Foreign Exchange Pool	Central Banks Foreign Exchange Pool	Parallel Markets Foreign Exchange Pool
Exports — Regulated Channels of Distribution	Agriculture									X		
	Food processing											
	Textiles											
	Other Industry											
	Electricity											
	Construction											
	Oil									X		
	Transportation & Comm.									X		
	Services											
Imports — Unregulated Channels of Distribution	Agriculture	X	X	X	X	X	X	X	X			
	Food processing	X	X	X	X	X	X	X	X			
	Textiles	X	X									
	Other Industry	X	X									
	Electricity											
	Construction											
	Oil	X	X					X	X			
	Transportation & Communica.	X	X	X	X	X	X	X	X			
	Services	X	X					X	X			
Imports — Regulated Channels of Distribution	Agriculture	X	X	X	X	X		X	X			
	Food processing	X	X	X	X							
	Textiles											
	Other Industry	X	X									
	Electricity											
	Construction											
	Oil											
	Transportation & Communica.											
	Services											

perfectly elastic. There is unemployment. In rural areas, labor supply is fixed and the wage adjusts to clear the market. However, all production sectors, whether in urban or rural areas are not assumed to pay the same wage. Public sector firms pay a given wage, considered a policy variable. The differentials between public sector wages and the basic rural and urban wages adjust. The differentials between the private sector wages and the basic wages are assumed fixed.

The reference model of the MISR2 class captures another feature of labor markets in Egypt: public sector activities are expected by the government to absorb a certain amount of the increase in the labor force. Thus state enterprises have a very limited leeway in their hiring decisions. Furthermore they cannot fire labor except under exceptional circumstances. For all practical purposes, their wage bill is fixed. This feature of employment policies is captured by assuming that labor, in public sector activities, is a fixed factor. This implies that a rent accrues to labor, a rent which may fall short or exceed the cash the public sector activities pay for labor services. The difference between the rent and the cash payments is absorbed by the operating surplus of these activities. Practically this amounts to a tax imposed on the public sector production activities.

Capital and land are the two other primary factors used in production. Land appears explicitly in agriculture and is assumed fixed within a period. It thus earns a rent which is channeled to the owners of the land. The treatment of capital is identical. Capital is immobile within a period and earns also a rent again distributed across owners.

3. Foreign Exchange and Trade

Trade with the rest of the world is an area where governments tend to get heavily involved. The MISR2 reference model captures the trade regime in Egypt by assuming three balance of payments with three exchange rates.

The first balance of payments records transactions through the central bank pool of foreign exchange. The exchange rate is nominally fixed and the market of foreign exchange clears through net transfers to the second balance of payments. Thus the first balance of payments works like a traditional foreign exchange market with a fixed exchange rate (price) and adjustment in net borrowing (quantity adjustment). The only difference is that the "net borrowing" is not from the rest of the world but from the second pool of foreign exchange.[1]

This second pool of foreign exchange finances transactions recorded in the second balance of payments. Again the nominal exchange rate is fixed, and the supply and demand of foreign exchange clears via rationing of imports. Once compulsory transfers to abroad are made, there is a certain amount of foreign exchange left. This is then allocated, in a policy decision, across various commodities. This allocation, jointly with the world

---

[1] The net borrowing from abroad through the first balance of payments is fixed and considered a policy variable.

prices and the exchange rate, determines the amounts of imports of each commodity allowed through this second pool.

The third balance of payments corresponds to the "parallel" market of foreign exchange. There the nominal exchange rate adjusts to equilibrate inflows and outflows of foreign exchange.

An important issue implied by the existence of three balance of payments is the allocation of imports between them. The last 18 columns of Table 3.1 distinguish between imports through unregulated and regulated channels of distribution. The Table 3.1 also shows the sources of these imports. Imports traded through regulated channels of distribution come entirely through the first balance of payments. Most of imports traded through unregulated channels can come through either of the three balance of payments. Consider, as an example, imports of "other-industry" goods. The goods coming through the three balance of payments are not identical; they are similar, substitutable. Depending on relative prices, between the three pools, traders will allocate their total demand for industrial goods imports by minimizing their costs. They will thus have a demand on each of the pools. However on the central and commercial banks pools, traders will be rationed. In order to obtain additional imports, they will need to go to the "parallel market" pool. This will drive up the price of foreign exchange on that market and push up the cost of imports. Wherever imports are rationed a rent larger than the cash payment will appear. The difference (between the rent and the cash payment) is an import premium which is channelled back to the final users of imports: households, companies.

4. The Macroeconomic Equilibrium

In MISR2, two savings pools and three categories of investors are identified. The two savings pools capture the public-private sector distinction in savings mobilization. The first category of investors correspond to production activities in the private sector while the second category corresponds to those in the public sector. The last category is the government.

The private savings pool collects the savings of all private agents in the economy, namely households and private companies. It also receives foreign savings coming through the parallel market of foreign exchange. The foreign savings are asumed fixed in dollars reflecting the view that their supply is inelastic. This is consistent with the flexibility of the exchange rate on the parallel market of foreign exchange. Total resources available to the private savings pool are then allocated to the public sector savings pool and to private investors. The allocation captures the issuing of bonds by the public sector in order to mobilize private resources. This process determines the availability of financial resources to private investors, constraining the total amount of investment they can undertake. The question of the allocation, between the various private sector investors, of the total amount of funds available to them, however, remains. It is dealt with along the lines proposedin Dervis, de Melo and Robinson (1982). In essence each private investor will see his share of funds increase if the rate of return of his activity is higher than the average rate of return in the private sector. Otherwise the share will decrease. The variation in shares takes place between periods; within a period they are kept constant.

The public savings pool receives funds from three sources: (i) public sector companies and government; (ii) the private sector, through the private sector savings pool and (iii) the rest of the world. Funds from the rest of the world are foreign savings coming through the central bank and the commerical banks foreign exchange pools. In both cases these foreign savings are assumed to be limited in amount. In addition to the view that they may not be forthcoming, authorities may not want to accelerate the accumulation of foreign debt. The public sector as a whole then faces a "self-imposed" constraint on the availability of foreign savings.[1] The limitation in the amount of foreign savings, together with doemstic funds, determines the total resources available to the public sector savings pool. On the other hand, authorities will decide on the total public sector investment budget. That is nominal investment will be fixed. The mechanism through which the public sector savings gap will be closed is the contraction or expansion of the economy. Indeed, the nominal value of the domestically generated savings [2] going to the public sector pool will need to increase in order to validate the nominal value of investment. In the process, because of resource constraints, including foreign exchange, the real value of nominally fixed variables be cut and in particular real public investment will adjust.[3]

One question remains to be clarified: what sets the general price level (GPL)? In practice a GPL is an aggregate price obtained as an index of several more specific prices. In the MISR2 reference model, several prices are fixed within the period, among them are the prices of public sector outputs, the prices of rationed goods, the urban wage, the world prices and the central and commercial banks exchange rates. All these prices tie down, to a certain extent, the GPL. Other prices in MISR2 adjust to clear markets. They ultimately respond to basic constraints in the economy in terms of production capacity, rural labor supply, foreign exchange availability. The tighter the latter, the larger aggregate demand and the higher the flexible prices will go relatively to the fixed ones. As a consequence any average price, hence a measure of the GPL, will reflect the way relative price changes affects the components of aggregate demand. An aggregate excess demand hits basic constraints forcing adjustments in aggregate demands and in supplies. What actually happens are shifts in relative prices against the commodities with fixed prices and in favor of the commodities which are in limited supply.

1/ This reflects policy makers concern about the accumulation of foreign debt.

2/ Consider the following relation $\overline{PI.I} = S + e\overline{F}$ where $\overline{PI.I}$ is the fixed value of public investment, e the fixed exchange rate, F fixed foreign savings and S domestically generated savings going to the public sector. If the left-hand side is larger than the right-hand side, S will need to increase to validate the equality. With given savings rate ($S = \alpha Y$ where Y is income) Y will expand. But because of resource constraints relative prices will change and affect the various categories of aggregate demand. In particular private sector prices and the "parallel" market exchange rate will rise relatively to the fixed prices in the economy: public sector prices and world prices.

3/ This is the process of adjustment which seems to best describe macroeconomic adjustment in the first part of the eighties.

CHAPTER 2: The Formulation of the MISR2 Reference Model

The previous chapter provided a non-technical presentation of the economics of the reference MISR2 model. It dealt with the equilibrium obtained in each period. This chapter provides a detailed documentation of the within-period module of the model. The chapter is organized into three parts. Part A gives a brief review of the transaction values (TV) approach which is used to implement the model. Part B presents the accounting framework (SAM) of the MISR2 reference model, while part C presents the TV specifications used in MISR2 and the within-period module of the reference model as formulated in TV.

A. The TV Approach

1. An Overview of TV

The within-period module is formulated and implemented along the Transactions-Values (TV) approach - see Drud, Grais, Pyatt (1983). Because of the presumed unfamiliarity of the reader with the approach and because it motivates the way the presentation of the module is given, we overview it here briefly.

The formulation of an economywide model requires three sets of information: (i) an accounting framework; (ii) a specification of behaviors and technologies; and (iii) constraints ensuring the consistency of the independent decisions of the agents intervening in the economy.[1] The TV approach organizes these informations in a systematic way using a Social Accounting Matrix (SAM). The SAM of the MISR2 is presented in detail in part B of this chapter.

A SAM provides the accounting of an economy in a matrix format where each agent has a row and a colum registering his receipts and outlays respectively. Thus each cell in a SAM represents a payment from a column to a row which is, in all generality the value of a transaction. All entries in a SAM can therefore be regarded as representing Transactions Values (TV's). The accounting framework of economywide models, specified along the TV approach, is provided by a SAM.

Going beyond the accounting framework the TV's reflect underlying agent behaviors and technologies. Consumer expenditures result from utility-maximization, factor payments from profit-maximization: all sorts of allocation rules may be envisaged. Hence an economywide model involves the modeling of the entries of a SAM through explaining the behavior underlying the realized TV's. The specification of the behavior associated with each

---

[1] In order to implement the model one needs also numerical values for the parameters and exogenous variables.

cell of the SAM will provide an almost complete information on behaviors and technologies. [1]

Having told how each agent intervening in the economy takes his decisions, one question remains: what ensures the consistency of all independent decisions? A consistent equilibrium is achieved for some vector of prices, quantities and values which makes all accounts in the SAM balance. The balancing of each account will rely either on price, quantity, value or on all three types of adjustments. Defining for each account which of the price, quantity or value attached to it adjusts will specify how all allocation decisions are made consistent and define equilibrium.

The TV approach has led to the development of a software for implementing and solving economy-wide models formulated along the lines described above. It requires from the user: (i) a SAM for a base-year; (ii) the TV specification of each cell; (iii) the system constraints indicating how each account balances. This information [2] is then used to solve for a general equilibrium under alternative assumptions on the policy parameteres.

In the following we first present, in part B, the accounting framework (SAM) of the reference model in the MISR2 class.[3] Part C of this chapter then gives a documentation of the TV specifications--the behaviors and technologies--and of the systems constraints which ensure the consistency of the independent decisions of the agents intervening in the economy.

B. The MISR2 SAM

1. An Introduction to the SAM Framework

The accounting and disaggregation of the MISR2 models are captured by an underlying Social Accounting Matrix (SAM). The first historic attempt to provide national accounting information in a "SAM" was developed for England for 1688 by Gregory King (Pyatt and Thorbecke 1976, pp. 22-23). The basic idea King developed has been extended in the United Nations System of National Accounts (the SNA) (UNSO 1968). However, the SNA emphasizes the structure of

---

[1] Each cell specification does not however provide a full account of behaviors and technologies. Indeed it will generally include parameters which values characterize behaviors and technologies at particular points in time and space. Any implementation requires a full information on the specifications of each cell and necessitates also numerical values for behavioral and technological parameters. An advantage of the TV approach is that most of these parameters can be derived from the base year SAM, in such a way that the model reproduces the base year data. Other parameters must however be obtained from other sources.

[2] In addition, the values taken by parameters, which cannot be derived from the SAM will also be needed.

[3] This is the model used to derive the reference path. The SAM is presented in considerable detail because understanding of this SAM helps to a great extent to understand the model.

production in an economy, largely disregarding distributional issues. The recent development of SAMs by Pyatt and associates (Pyatt and Roe 1977) has concentrated on flows of incomes and expenditures between the major institutional participants in an economy, attempting to capture income distribution.

In essence, a SAM is a consistent data system that provides comprehensive base-year information on such variables as: (i) the structure, composition and level of production; (ii) the factoral value added; and (iii) the distribution of income among household groups. Typically, a SAM is organized around an input-output table, and includes information on consumption and production patterns, exports, imports, investment, and savings. Depending on the particular issues of interest and the data available, a SAM may include more detailed information on income distribution, tax structure, and monetary variables. The most important feature of a social accounting matrix is that it provides a consistent, convenient approach to organizing economic data for a country, and it can provide a basis for the formulation of economywide frameworks for policy analysis.

The conceptual framework of a SAM is explained in Pyatt and Thorbecke (1976) and King (1981). Several well documented SAMs have been compiled during the last decade and work on SAMs is progressing in many countries. All of these SAMs are "accounting" SAMs, since they are focused on the one hand mainly on organization, and reconciliation of statistical data, and on the other hand on the descriptive analysis of economywide data for a particular country.

Recently a new generation of "modeling" SAMs and the Transaction Value (TV) approach have been developed. These SAMs serve as data base, accounting base, and algebraic statements of economywide equilibrium models. The TV approach is a systematic and flexible way of formulating and implementing economywide multisectoral equilibrium models based on a SAM (Drud, Grais, and Pyatt 1983). The first attempt at using the TV approach was made for the SIAM1 model of Thailand (Drud, Grais, and Vujovic 1982). Most recently, the Thailand SAM of the SIAM 2 model (Amranand and Grais, 1983) has been contructed as the most disaggregated and comprehensive "modeling" SAM (Chewakrengkai and Lamsam 1982). Simultaneously with the SIAM 2 model for Thailand, the MISR 2 model for Egypt has been developed. Both of these models use the TV approach. They are the latest vintage of economywide frameworks for policy analysis. The major advantages of the approach used in these models is that it allows a large flexibility in the specification of the model. Alternative behaviors, policy regimes and institutional arrangements can be captured in a fairly simple way.

The SAM underlying the MISR 2 models of Egypt is derived from the 1979 SAM 2 for Egypt built by the DRTPC. [1]

In the following the configuration of the accounts of the MISR2 models and their underlying SAM are explained. This part of the chapter is organized into three sections. Section one describes an aggregate SAM for Egypt. Section two describes an intermediate disaggregation of the accounts focusing on the main economic issues the MISR2 models are designed to address. These issues are pricing and trade policies. The emphasis of the intermediate disaggregation is on the description of distinctions between on one hand activities and commodities and on the other hand of the rest of the world accounts which are the relevant accounts for the issues considered. The third section describes the fully disaggregated SAM on which the MISR2 models are based.

2. The Aggregate SAM for Egypt

The aggregate SAM for Egypt is presented in Table A.1.a. It consists of six basic accounts: (i) factors of production; (ii) currrent accounts of institutions; (iii) consolidated capital accounts; (iv) activities; (v) commodities; and (vi) the rest of the world. This macroeconomic SAM for Egypt for the year 1979 presents the aggregate income and expenditures flows between the six categories of accounts.

The first class of accounts in Table A.1.a, Factors of Production, identify the receipts and disbursements of factor incomes within the economy. Factor income is derived from employment in government (LE 1242.8 million), from domestic production activities (intersection of column eight with rows 1, 2, and 3), and from abroad. This is shown in first three rows, while the allocation of this factor income between domestic institutions, and payments abroad of factor income, are shown in first three columns. Three factors of production are distinguished in this aggregate SAM for Egypt: (1) labor, (2) capital, and (3) land. Row account totals for each factor give detailed view of the functional, or factoral, distribution of income by source within the economy, while the columns of the factor account indicate who receives these factor incomes. Factors of production are further disaggregated in the detailed accounts which are presented in the following sections.

The second class of accounts in Table A.1.a, the Current Institutions

---

[1] Development Research and Technological Planning Center of Cairo University. Concepts underlying SAM 2 are summarized in Pleskovic and Crosswell (1981); for a detailed documentation of the SAM 2, see: Working Paper No. 7, "SAM 2 and Documentation," DRTPC, Cairo University, 1982. This SAM 2 for Egypt is a part of the output of a larger joint project of the DRTPC, Cairo University, U.S. Agency for International Development (AID) and the World Bank. Several Egyptian agencies and professionals participated in compiling the SAM 2 for Egypt, including representatives of DRTPC, Ministry of Planning, CAPMAS (Central Agency for Public Mobilization and Statistics), and Institute of Planning. The development of the MISR 2 model and its SAM would not have been possible without their valuable efforts and participation.

TABLE A.1.a
Aggregate SAM for Egypt -- SAM1
(Millions of L.E.)

			I FACTORS OF PRODUCTION			II INSTITUTIONS CURRENT ACCOUNT					III CONSOLIDATED CAPITAL ACCOUNT	IV ACTIVITIES	V COMMODITIES	VI REST OF WORLD		VII
			LABOR	CAPITAL	LAND	HOUSEHOLDS	PRIVATE COMPANIES	PUBLIC COMPANIES EGPC	PUBLIC COMPANIES OTHER	GOVERNMENT				FOREIGN EXCHANGE BUDGET	OWN EXCHANGE	TOTAL
			1	2	3	4	5	6	7	8	9	10	11	12	13	14
I FACTORS OF PRODUCTION	LABOR	1										2872.6		666.2	883.3	5664.9
	CAPITAL	2										7635.9		41.6		7677.5
	LAND	3										316.7				316.7
II INSTITUTIONS CURRENT ACCOUNT	HOUSEHOLDS	4	5578.4	3586.6	306.5		14.9	9.6	166.6	433.2				158.9		10254.7
	PRIVATE COMPANIES	5		412.8		1.0			0.8							414.6
	PUBLIC COMPANIES EGPC	6		1637.4					99.8	77.5						1814.7
	PUBLIC COMPANIES OTHER	7		2013.7	10.2	29.2	3.2	58.0		489.0				3.8		2607.1
	GOVERNMENT	8		27.0		766.7	26.3	1230.6	1073.6			192.6	411.1	104.6		3832.5
III CONSOLIDATED CAPITAL ACCOUNT		9				1033.4	370.2	109.0	975.1	600.5				1061.8		4150.0
IV ACTIVITIES		10											21080.8			21080.8
V COMMODITIES		11				8222.0				837.0	4150.0	10063.0		3743.3		27015.3
VI REST OF WORLD	FOREIGN EXCHANGE BUDGET	12	86.5			202.4		407.5	291.2	152.5			4640.1			5780.2
	OWN EXCHANGE	13											883.3			883.3
VII	TOTAL	14	5664.9	7677.5	316.7	10254.7	414.6	1814.7	2607.1	3832.5	4150.0	21080.8	27015.3	5780.2	883.3	

Source: Working Paper No. 2, "SAM1 and Documentation," DRTPC, Cairo University, 1982.

Accounts, divide institutions into five categories: households, private companies, public companies, and government. Households and companies are domestic institutions receiving factor income. For example, households receive LE 5578.4 million from labor, LE 3586.6 million from capital, and LE 306.5 million from land rent. Domestic institutions also receive transfers and spend part of their income as transfer payments. Government income collected in row 8 is almost entirely tax and transfer income with the exception of imputed rent on government dwellings. On the other hand, this income is spent on transfers to households, companies, and the rest of the world, or it is spent on commodities as government consumption (i.e., LE 837.0 million). Companies distribute part of their profits to households in form of dividends. The sums of all receipts of factoral income and transfers by institutions represent the distribution of income among designated institutions.

The third main account of the aggregate SAM for Egypt is the consolidated capital account. Receipts along the row show savings (transfers from currrent to capital account) of each domestic institution, together with current transfers from abroad. The row sum of this account (LE 4150.0 million) provides the necessary finance for gross domestic fixed capital formation and changes in stocks.

Next two accounts are for production activities and for commodities they produce. The column entries of the production account indicate the costs incurred in production, including payments for services of factors of production, payments for intermediate inputs (LE 10063.0 million); and paymenst of taxes (fees and licences). The sum of this column gives domestic production valued at "factory gate" prices. The output of the domestic production activity (LE 21080.8 million) appears at the intersection beween commodity column and activity account row. A second entry in the commodity column is net indirect taxes (LE 411.1 million). This entry represents the difference between aggregate domestic production valued at factory-gate prices and at market prices. Next two entries of the commodity column represent imports at official and own exchange rates, respectively. Together, these entries sum to the total commodity supply valued at market prices.

Commodities are distributed to satisfy domestic final demand, intermediate demand for production, and export demand along the row of the commodity account. Final demand consists of household consumption LE 8,222.0 million, public consumption LE 837.0 million, and investment LE 4,150.0 million; intermediate demand is equal to LE 10,063.0 million; and export demand amounts to 3,743.3 million.

The final account of the aggregate SAM for Egypt, Table A.1.a, is the rest of the world account. This account is subdivided into the foreign exchange budget and the own exchange account. The former reflects transactions that are undertaken at the official exchange rate. The latter row and column, on the other hand, account for imports which are financed through savings of the Egyptians working abroad and which are bought at the "market" exchange or own exchange rate.

3. An Intermediate Disaggregation of the SAM Accounts

The SAM in Table A.1.a is the most aggregated version of the data base for the MISR2 model. However, the accounts of the SAM underlying the

MISR2 model are disaggregated to a larger extent, providing a final SAM composed of 362 x 362 accounts. Because it is impossible to present such a complete matrix on one page, two steps are taken to provide the basic understanding of the final SAM and interactions between the individual accounts. The first step presented below is a description of two intermediate SAMs, each focusing on the main economic issues that underlie the accounting structure of the model. One of these intermediate SAMs, Table A.1.b, outlines the major groups and features of the production activities and commodities accounts. The other, Table A.1.c, describes the interactions of the rest of the world accounts with other aggregate accounts.

All of these accounts are then disaggregated in detail, one by one in the second step (Tables A.1.1 to A.1.19). Disaggregated individual accounts are presented in the third section of this chapter as independent medium size matrices. These are, however, too large to be composed into one final SAM matrix. In order to find out how these disaggregated matrices fit into the final SAM it is necessary to use the two intermediate SAMs as a reference point. These two SAMs therefore show the major interactions between the more aggregate accounts, whose internal structure can be found in the tables (matrices) presented in the third part of this paper.

3.1. An Intermediate Disaggregation of Activities and Commodities

Table A.1.b presents an intermediate level of disaggregation of the commodities and activities accounts of Table A.1.a. A new account for composite inputs (intermediates) is also introduced in this table (columns/rows 2 to 5), while factors of production and the rest of the world account are aggregated each into one row and column. Otherwise Table A.1.b keeps the same disaggregation as Table A.1.a. Its purpose is to present the main structure of disaggregated commodities and activities accounts of the SAM underlying the MISR2 model.

One of the major emphasis of the MISR2 model and its SAM is on the analysis of pricing policies in the Egyptian economy. The structure of pricing policies is captured in the SAM framework through disaggregation of activities into public and private activities; and through disaggregation of commodities into government trade and non-government trade commodities both subdivided into domestic, imported, and exported commodities.

In Table A.1.b activities are disaggregated into public and private activities accounts. Each of these consist of nine sectors described in detail in the main paper. One of the key features of the production account for Egypt is the distinction between public and private activities. This division has been made to reflect the major presence of the public sector in production, accounting for over fifty percent of the gross output. The rationale for the distinction between public and private activities is that producers may behave differently in response to different incentives present in each of the two sectors. For instance, public production in Egypt is regulated to a great extent in terms of the quantity to be produced; the price received for the output; the level of employment and wages; and taxation or transfers of operating surplus to the government. On the other hand, private sector production corresponds more closely to traditional profit-maximizing behavior, although prices and quantities are regulated in some cases (e.g., cotton). A second rationale is to be able to examine differences between the structure of production and value added in the public and private sectors.

TABLE A.1.b
Intermediate SAM for Egypt -- Disaggregation of Activities and Commodities

			Composite Inputs				Current Institutions					Activities		Commodities — Govt. Trade			Commodities — Non-Government Trade								
			Domestic		Imported												Domestic		Exports		Composite		Imports			
			Public Activities	Private Activities	Public Activities	Private Activities	Households	Private Companies	Public Companies	Government	Rest of the World	Capital Account	Public	Private	Domestic	Imported	Exported	Public	Private	Public	Private	Domestic	Exports	Central Bank	Commercial Banks Pool	Own Exchange	Composite Imports
			1	2	3	4	5	6	7	8	9	10	11	12	13	14	15	16	17	18	19	20	21	22	23	24	25	26

(Schematic SAM matrix with circles indicating non-zero entries; row labels: Primary Factors (1); Composite Inputs — Domestic: Public Activ. (2), Private Activ. (3); Imported: Public Activ. (4), Private Activ. (5); Current Institutions: Households (6), Private Companies (7), Public Companies (8), Government (9), Rest of the World (10); Capital Account (11); Activities: Public (12), Private (13); Commodities — Govt. Trade: Domestic (14), Imported (15), Exported (16); Commodities — Non-Government Trade: Domestic Public (17), Domestic Private (18), Exports Public (19), Exports Private (20), Composite Domestic (21), Composite Exports (22), Imports Central Bank (23), Commercial Bank Pool (24), Own Exchange (25), Composite Imports (26).)

- 123 -

For example, value added generated in private activities exceeds the value added by public activities; public sector production is concentrated in specific economic sectors and reveals distinct patterns of exports, imports, and intermediate consumption.

Public and private activities appear in columns/rows 12 and 13 in Table A.1.b. In this SAM columns 12 and 13 show the cost payments of activities to primary factors (labor, capital, and land), to composite inputs (domestic and imported intermediate inputs), and to the government (fees and licences). The sums of these two columns represent the aggregate production of public and private activities (at producers cost). These productions are supplied to the commodity accounts for domestic use and exports in rows 12 and 13.

The commodity accounts, presented in columns/rows 14 to 26, are subdivided into several subaccounts, because in the Egypt economy some activities are assumed to produce several commodities and some commodities are produced by different activities. The intersection of rows 12 and 13 with columns 14 to 26 represents the mapping between activities and commodities. Commodity accounts buy the outputs of public and private activities and pay the commodity taxes to the government in row 9. The sums of commodity columns represent the supplies of domestic goods and services at market prices.

Table A.1.b distinguishes two major groups of commodities: (1) government trade commodities, and (2) non-government trade commodities. Government trade [1] is a special institution created by the Egyptian government. The major role of this institution is to buy domestic or imported goods and to deliver them to the distribution companies (wholesale trade), directly to consumers and producers or the the rest of the world (exports). The government trade also finances the difference between purchasing prices and selling prices at which the goods are delivered to consumers. The difference represents either subsidies to consumers or producers who buy government traded goods; or export taxes in case of government trade exports, since these are bought at a lower price on the domestic market and sold at world prices to the rest of the world. Thus, the distribution of commodities through government trade channels is the main vehicle for implementing subsidies or indirect taxes in Egypt.

As indicated in Table A.1.b, in principle each of the goods and services can be distributed through government trade channels, through non-government trade channels, or through both. Within each distribution channel there are distinctions among imports, exports, and domestically produced goods and services (both public and private) destined for the domestic market or for exports. As goods and services flow from production activities into these various commodity accounts, their value (price) is altered to reflect distribution margins, indirect taxes, subsidies, import tariffs, or export taxes.

For instance, in column 14, domestic government trade commodities are bought from public and private activities from rows 12 and 13, while indirect taxes on these goods and services appear at the intersection of the government

---

[1] For a detailed description of government trade, see Working Paper No. 7, "Sam 2 and Documentation, Government Trade Sector," DRTPC, Cairo University, 1982.

account (row 9) and column 14. The sum of column 14 thus measures supply of domestic government commodities at market (or user's) prices. Similarly, government trade imports appear at the intersection of column 15 with row 10 (rest of the world), while import tariffs appear in row 9. Government trade exports are measured in the same way. They are bought from the activity account in column 16, where export taxes appear in the government account in row 9.

Table A.1.b also introduces composite accounts for non-government commodities and intermediate inputs. Composite account is a convenient account introduced for goods that are more or less perfectly substitutable. Three composite accounts are distinguished in Table A.1.b. First of these is presented in columns 2 to 5 (composite inputs), where commodities are a composition of government trade and non-government trade domestic and imported commodities. The second composite account is indicated in columns 21 and 22, where commodities are a combination of public and private commodities. The third composite account is for imports. Composite imports are indicated in column 26. These are a combination of the Central Bank, commercial banks, and own exchange imports. All of the composite accounts are explained in more detail below as a part of the structure of other SAM accounts.

Non-government trade commodities consist of domestic, exported, and imported commodities in the same way as government trade commodities. The only difference between the two accounts is that non-government trade commodities are traded through the composite accounts. That means, for example, that non-government trade domestic commodities are at first bought in columns 17 and 18 from public and private activities (rows 12 and 13). These commodities are then sold from rows 17 and 18 to the composite account at the intersection with column 21, where indirect taxes appear in the government account in row 9. The sum of the three entries in column 21 gives the supply of non-government domestic composite commodities at market prices. Non-government trade exports are treated in the same way, except that export taxes appear in row 8, public companies (EGPC), instead of in row 9.

Non-government trade imports are treated in a similar way, except that these imports are traded through three different channels. These three channels are (1) central bank pool, (2) commercial banks pool, and (3) own exchange pool. The central bank and commercial banks pools represent imports bought at two different official foreign exchange rates, while own exchange pool represents a parallel market. These three types of imports appear in the SAM at the intersection of row 10 with columns 23, 24, and 25. These imports are then in turn bought by the composite imports account (column 26) from rows 23, 24, and 25. Import tariffs appear at the intersection of row 9 with column 26, where the sum of column 26 represents the supply of non-government imports at market prices. The entry in the cell 24/24 indicates import premia which is due to rationing. Non-government imports are explained in more detail in the next sections (Table A.1.c and Table A.1.18).

The columns of the government trade commodity accounts and non-government trade composite accounts give the price structure and (implicitly) the rate of indirect taxes and subsidies. Rows of these commodity accounts, on the other hand, represent the supply of commodities to meet demand, i.e., consumption of commodities at market prices by households, government, capital account, rest of the world, and composite inputs account (intermediate demand). Goods and services are distributed directly from the two distribution channels to final users. In other words, domestic and foreign demand by private consumers, investors, the government, production activities, and the rest of the world is disaggregated according to the distribution channel (government and non-government trade) and origin (domestically produced or imported) for each of the 9 goods and services. This type of disaggregation of the final and intermediate demand allows for distinction between consumption of government and non-government traded domestic, imported and exported commodities. Since each of these commodities is taxed or subsidized at different rates, commodity accounts represent different pricing policies (commodity markets) in Egypt. In this sense, commodity accounts serve as an accounting basis for economic modeling with respect to pricing and income distribution issues.

As noted above, Table A.1.b introduces the composite inputs account. The purpose of this account is to capture the accounting structure and the composition of intermediate inputs bought by activities. For example, intermediate inputs used by public activities are bought in column 2 in Table A.1.b from row 14 and row 21. These two entries indicate government trade domestic and non-government trade composite domestic commodities. There is an additional entry at the intersection of column 2 with primary factors account in row 1. This entry indicates a rent which occurs to capital because government trade commodities are rationed and subsidized. Other three columns (3 to 5) are treated in the same way as column 2. The sums of columns 2 to 5 represent total domestic and imported intermediate inputs. These are sold to public and private activities from rows 2 to 5 at the intersection with columns 12 and 13.

This description of the commodity accounts in Table A.1.b exhausts the explanation of interactions between commodity and activity accounts. As indicated above, with the exception of the composite inputs account and disaggregated activity/commodity accounts, all other interactions in Table A.1.b follow the same accounting structure as presented in Table A.1.a.

3.2. An Intermediate Disaggregation of the Rest of the World Accounts

Table A.1.c represents an intermediate disaggregation of the rest of the world accounts and their interactions with other aggregate accounts. The major feature of the disaggregated rest of the world accounts is that they attempt to capture foreign trade regime in Egypt. There are currently three different exchange rates operating in Egypt: (1) the official (Central Bank) rate, (2) the commercial banks rate, and (3) the own exchange rate. The official rate applies to the Central Bank pool which is supplied by receipts from Suez Canal dues, petroleum and some agricultural exports. Central Bank funds are used primarily to service external government debt and to finance imports of basic supply commodities.

The commercial banks pool is supplied mainly by cash remittances from Egyptians working abroad, tourism receipts and non-oil and cotton export receipts. Commercial banks funds are used to finance the foreign trade of public sector companies that are included in the foreign exchange budget but not financed by the Central Bank pool. The commercial banks exchange rate is different from the official (Central Bank) rate, being in between the official rate and the own exchange market.

The "free" or own exchange market is supplied by remittances from Egyptians working abroad and some tourism receipts. Private individuals and companies can buy foreign exchange at a freely floating rate in order to finance imports or to invest in dollar-denominated assets.

TABLE A.1.c

Intermediate SAM for Egypt — Disaggregation of the Rest of the World

Therefore, there are three major channels through which exchange with the rest of the world takes place in Egypt: (1) Central Bank pool, (2) commercial banks pool, and (3) own exchange pool. These three channels and transfers between them are captured in columns/rows 7 to 12 in Table A.1.c. The rest of the world thus consists of six accounts: (i) Central Bank, (ii) transfer account, (iii) commercial banks, (iv) own exchange, (v) pool import premia, and (vi) discretionary foreign exchange budget. The intermediate SAM in Table A.1.c is similar to the SAM presented in Table A.1.b with two exceptions. One exception is that current institutions accounts: households, private companies, public companies, and government are aggregated into one row and one column to keep the SAM smaller. The other exception is that the rest of the world account comprises of six accounts instead of one as in Table A.1.b. Activities and commodities accounts are kept disaggregated in Table A.1.c because the rest of the world accounts interact to a large extent with these accounts, especially with imported commodities. The rest of the accounts in Table A.1.c are the same as in the intermediate SAM in Table A.1.b.

In the following each column and row of the rest of the world account is explained in terms of expenditures and receipts of these accounts. For example, column 7 of the intermediate SAM indicates foreign exchange outflows of the Central Bank pool, while row 7 represents foreign exchange receipts of the Central Bank pool. Specifically, the intersection of row 1 with column 7 represents capital earnings (interest) received by factors from abroad; row 6 shows a value of transfers to current institutions from the rest of the world; row 13 shows foreign savings plus changes in reserves (or foreign account deficit); row 18 indicates that all government trade exports are traded through the Central Bank pool; and row 24 shows that a portion of non-government exports is traded through the Central Bank ($T_{24.7}$) and the remaining part through the commercial banks pool ($T_{24.9}$).

Foreign exchange receipts of the Central Bank pool can be read along row 7 of the SAM. The intersection of this row with column 6 represents a portion of current institutions transfers abroad that are channeled through the Central Bank. The entry in row 7 and column 8 indicates excess of foreign exchange on the Central Bank pool being transferred to the commercial banks. And the entry in row 7, column 17 indicates payments for government trade imports to the Central Bank pool.

The next account (column 8) represents transfers between the Central Bank pool and the commercial banks pool, where $T_{7.8}$ has been defined above, while $T_{11.8}$ represents a difference between the Central Bank exchange rate and the commercial banks exchange rate (pool import premia). The sum of column 8 is transfered from row 8 to the commercial banks pool at the commercial banks exchange rate.

Outflows of the commercial banks pool are indicated in column 9 and consist of factors and transfer income (rows 1 and 6), transfer income from the Central Bank (row 8), foreign account deficit of the commercial banks pool (row 13), and expenditures on non-government trade exports (row 24). On the other hand, receipts of the commercial banks pool (row 9) consist of factor and transfer incomes sent abroad (columns 1 and 6), and total foreign exchange availability for discretionary foreign exchange budget received from column 12.

Outflows from the own exchange budget are indicated in column 10. They consist of factor income (row 1), and foreign savings plus changes in reserves (row 13). Receipts of the own exchange budget are from own exchange imports which appear at the intersection of row 10 with column 28.

Expenditures of the next account, pool import premia (column 11), consist of premia paid to households and companies. This premia occurs to current institutions due to differences between the Central Bank and commercial banks exchange rates. Receipts of this account appear in row 11 and are derived from the transfer account (column 8) and non-government imports which appear in columns 25 and 27 in Table A.1.c.

The last account of the rest of the world is discretionary foreign exchange budget. Expenditures of this account are allocated to the commercial banks pool in row 9, while receipts of this account are derived from non-government imports which appear in column 26.

This description of the disaggregated Rest of the World account presents sources and uses of incomes of this account, as well as its major interactions with other SAM accounts.

4. Disaggregated Individual Accounts of the MISR2 Model and its SAM

The previous two sections described the aggregate SAM for Egypt and two intermediate SAMs with an attempt to highlight the major economic issues addressed by the MISR2 model and its accounting structure. The third section gives an additional presentation of the SAM accounts. The primary purpose of this presentation is to describe the accounting structure of the model in more detail, to present its data base, and the main interrelationships between the disaggregated accounts. The SAM tables presented in this section are described each separately following the major blocks of the aggregate SAM and two intermediate SAMs. In order to derive a complete picture of the disaggregated SAM, it is necessary to trace these tables back to the intermediate SAMs presented in Tables A.1.b and A.1.c.

4.1. Factors of Production

4.1.1 - Labor

Factors of production are aggregated into one row and one column in Table A.1.b. In the tables presented in this section, primary factors: labor, capital, and land are all further disaggregated. The primary purpose for a more detailed disaggregation of factors of production is a concern to monitor employment and income distribution implications of the policy instruments captured by the model.

Tables A.1.1 and A.1.2 represent disaggregation of urban and rural labor, respectively. Urban labor is employed in the public and private sectors and the government. For example, wages paid to urban labor by the government, state enterprises (public activities), and private enterprises (private activities) appear at the intersection of rows 2 and 3 with columns 5 to 21 in Table A.1.1. Sectoral wage differentials for a particular activity appear in row 4, i.e., revenue account of urban households. The sums of columns 5 to 21 thus give average urban wages for specific public and private sectors.

TABLE A.1.1
URBAN LABOR

			PRIMARY FACTORS — URBAN LABOR			CONSOLIDATED ACCOUNTS				STATE ENTERPRISES									PRIVATE ENTERPRISES						
			1	2	3	4	5	6	7	8	9	10	11	12	13	14	15	16	17	18	19	20	21
			LABOR ABROAD	URBAN	PUBLIC URBAN	PRIVATE URBAN	GOVERNMENT EMPLOYMENT	AGRICULTURE	FOOD PROCESSING	TEXTILES	OTHER INDUSTRIES	ELECTRICITY	CONSTRUCTION	OIL	TRANSPORTATION AND COMMUNICATIONS	SERVICES	AGRICULTURE	FOOD PROCESSING	TEXTILES	OTHER INDUSTRIES	CONSTRUCTION	TRANSPORTATION AND COMMUNICATIONS	SERVICES
CONSOLIDATED ACCOUNT	PRIMARY FACTORS URBAN LABOR	1 URBAN			○	○																		
		2 PUBLIC URBAN					○	○	○	○	○	○	○	○	○	○							
		3 PRIVATE URBAN															○	○	○	○	○	○	○
CURRENT INSTITUTIONS ACCOUNTS		4 URBAN HSEHOLD REVENUE ACCOUNT	○	○			○	○	○	○	○	○	○	○	○	○	○	○	○	○	○	○	○
		5 REST OF WORLD 2 COMMERCIAL BANKS		○																			

TABLE A.1.2
RURAL LABOR

			PRIMARY FACTORS RURAL LABOR																			
				CONSOLIDATED ACCOUNTS				STATE ENTERPRISES									PRIVATE ENTERPRISES						
			LABOR ABROAD	RURAL	PUBLIC RURAL	PRIVATE RURAL	GOVERNMENT EMPLOYMENT	AGRICULTURE	FOOD PROCESSING	TEXTILES	OTHER INDUSTRIES	ELECTRICITY	CONSTRUCTION	OIL	TRANSPORTATION AND COMMUNICATIONS	SERVICES	AGRICULTURE	FOOD PROCESSING	TEXTILES	OTHER INDUSTRIES	CONSTRUCTION	TRANSPORTATION AND COMMUNICATIONS	SERVICES
			1	2	3	4	5	6	7	8	9	10	11	12	13	14	15	16	17	18	19	21	22
CUR-RENT INST. ACCT.	CONSOLIDATED ACCOUNT	RURAL	1			○	○																	
		PUBLIC RURAL	2					○	○	○	○	○	○	○	○	○	○							
		PRIVATE RURAL	3															○	○	○	○	○	○	○
PRIMARY FACTORS RURAL LABOR		RURAL HSEHOLD REVENUE ACCOUNT	4	○	○	○	○	○	○	○	○	○	○	○	○	○	○	○	○	○	○	○	○	○

The sums of rows 2 and 3 represent consolidated urban wages for aggregate public and private sector. Consolidated wages are treated in the column accounts in the same way as disaggregated activity wages. For example, the entry in cell 1.3 indicates total urban domestic wages paid by the public sector, the entry in cell 4.3 represents wage differential between public and private urban sectors, while the total of column 3 indicates the average wage in public sector.

Remittances received by urban labor are paid from the labor abroad (column 1) to urban household revenue account in row 4. Total urban labor is derived by summing up entries in row 4, and urban labor income transfered abroad is indicated at the intersection of row 5 with column 2. Table A.1.2 is the same as Table A.1.1, except that it represents disaggregation of rural labor.

Table A.1.3 represents a "make up" matrix for total urban and rural labor employed by public and private enterprises. For example, the sum of column 2 gives total labor employed by public agriculture, where entry in row 2 indicates urban agricultural labor and entry in row 19 rural agricultural labor. There is an additional entry at the intersection of column 2 with row 35. This entry represents a rent that occurs to capital because of controlled wages in public activities. Total labor employed by private enterprises is indicated in columns 11 to 17. The accounting structure of those columns is the same as for public enterprises, except that there is no entry for a capital rent because wages in the private sector are not controlled.

4.1.2 Capital and Land

Table A.1.4 describes a treatment of disaggregated capital and land accounts. As indicated in this table, the operating surplus of public and private activities is paid from columns 2 to 17 to urban and rural households in rows 3 and 4. This matrix reflects pattern of ownership of public and private sector companies. The capital incomes generated by various activities are on the whole distributed on the one hand to urban and rural households and on the other hand to companies. This is shown in columns 2 to 17 in Table A.1.4. The capital incomes accruing to households from public and private companies represent profit sharing and dividends. The capital incomes are treated in the same way for all activities with the exception of agriculture, electricity and oil. Because of relatively small size of the sectors, in agriculture and electricity (columns 2 and 7) all of the operating surpluses are allocated to public companies (row 7). In the case of oil, the EGPC has been separated from other public companies in order to differentiate between the role of the two accounts. Column 1 of Table A.1.4 represents national capital account. This account collects in its row net capital income from abroad (Table A.1.12) and rents which are due to rationing of domestic markets (fixed prices in some public companies, e.g., Table A.1.17). The total receipts of the national capital account are reallocated to households, companies and the government at the intersection of column 1 with rows 3 to 8.

Land incomes are captured in row 2 and columns 18 to 20. Consolidated rent is indicated in row 2 and is derived from public and private land in columns 19 and 20. The sum of row 2 is then allocated in column 18 to urban and rural households and other public companies at the intersection of rows 3, 4 and 7, respectively.

TABLE A.1.3
TOTAL LABOR

TABLE A.1.4
CAPITAL AND LAND

				PRIMARY FACTORS																		
						CAPITAL															LAND	
					PUBLIC									PRIVATE								
				NATIONAL CAPITAL	AGRICULTURE	FOOD PROCESSING	TEXTILES	OTHER INDUSTRIES	ELECTRICITY	CONSTRUCTION	OIL	TRANSPORT & COMMUNICATIONS	SERVICES	AGRICULTURE	FOOD PROCESSING	TEXTILES	OTHER INDUSTRIES	CONSTRUCTION	TRANSPORT & COMMUNICATIONS	SERVICES	CONSOLIDATED RENT	LAND 1 -- PUBLIC	LAND 2 -- PRIVATE
				1	2	3	4	5	6	7	8	9	10	11	12	13	14	15	16	17	18	19	20
CURRENT INSTITUTIONS ACCOUNT	CAPITAL	NATIONAL CAPITAL	1																				
	LAND	CONSOLIDATED RENT	2																			○	○
	HOUSE-HOLDS	URBAN REVENUE ACCOUNT	3	○	○	○	○	○		○		○	○	○	○	○	○	○	○	○	○		
		RURAL REVENUE ACCOUNT	4	○	○	○	○	○		○		○	○	○	○	○	○	○	○	○	○		
	PRIVATE COMPANIES		5	○										○	○	○	○	○	○	○	○		
	PUBLIC COM-PANIES	EGPC	6	○							○												
		OTHERS	7	○	○	○	○	○	○	○	○	○	○								○		
	GOVERNMENT	GOVT. REVENUE	8	○									○										
PRIMARY FACTORS																							

4.2 Composite Inputs

Table A.1.5 represents the structure of composite domestic intermediate inputs used by public activities. This table is in essence disaggregated column 2 of the intermediate SAM presented in Table A.1.b As it is clear from Table A.1.5, public activities buy domestic intermediate inputs from government trade channel and non-government trade channel. The accounting structure of this matrix is straightforward when intermediate inputs are bought from the private channel (non-government commodities) only. For example, public activity agriculture buys composite non-government commodities directly from nine sectors (intersection of column 1 with rows 2 to 10). This is the same for other activities which appear in columns 5 to 11.

However, a different accounting procedure is used for intermediates that are bought from both government trade and non-government trade channels. Because government trade commodities are both, rationed and subsidized, additional subaccounts are introduced to capture the difference in pricing between government traded and private commodities. For example, food processing activity is disaggregated into 3 subaccounts. In column 2 the food processing activity buys non-government intermediate commodities from rows 3 to 10, and a composite domestic commodity agriculture from row 12. Composite agriculture consists of both government trade and non-government trade agriculture as indicated in column 3 in cells 2.3 and 11.3. The entry in cell 11.3 is a rent on the rationed input delivered by the government. This rent is divided into two elements presented in column 4 in Table A.1.5. Cell 1.4 represents the cash paid to buy the agricultural commodity, and cell 13.4 indicates the difference between the actual cash paid and the total rent of the fixed factor. The latter appears in row 13 which is capital income of the food processing activity because it changes the capital income earned by this sector.

Table A.1.6 represents a disaggregation of the column 3 of the intermediate SAM presented in Table A.1.b. The accounting structure of this table is exactly the same as in Table A.1.5, except that intermediate inputs are bought by private activities instead of public activities.

Table A.1.7 represents a disaggregation of the column 4 of Table A.1.b. The accounting structure of this table is similar to the former two tables. This matrix represents the composition of imported intermediate inputs used by public activities. Intermediate imported inputs bought by public agriculture (columns 1 to 3) are treated in Table A.1.7 in the same way as the food processing activity in the previous two tables. The only difference between Table A.1.7 and the former two tables is that food processing activity buys composite imported inputs from both agriculture and food processing, government trade and non-government trade commodity accounts. For this reason, there are two rent accounts in columns 4 to 8, one for food processing and another for agriculture. In all other respects this table confirms exactly to the concept used in the previous two tables.

Table A.1.8 is a disaggregated column 5 of the intermediate SAM in Table A.1.b, representing composite imported intermediate inputs used by private activities. This table is in principle the same as the previous three tables of composite inputs.

TABLE A.1.5

COMPOSITE DOMESTIC INTERMEDIATE INPUTS: PUBLIC ACTIVITIES

				COMPOSITE DOMESTIC INTERMEDIATE INPUTS									
							PUBLIC ACTIVITIES						
				AGRICULTURE	TOTAL INTERMEDIATE	FOOD PROCESSING COMPOSITE AGRICULTURE	FOOD PROCESSING AGRICULTURE RENT ACCOUNT	TEXTILES	OTHER INDUSTRIES	ELECTRICITY	CONSTRUCTION	OIL	TRANSPORT AND COMMUNICATIONS	SERVICES
				1	2	3	4	5	6	7	8	9	10	11
NON-GOVERNMENT COMMODITIES	COMPOSITE DOMESTIC	AGRICULTURE	SUBSIDIZED VALUE	1	○				○	○	○	○	○	○	○
		AGRICULTURE		2		○			○	○	○	○	○	○	○
		FOOD PROCESSING		3		○			○	○	○	○	○	○	○
		TEXTILES		4		○			○	○	○	○	○	○	○
		OTHER INDUSTRIES		5		○			○	○	○	○	○	○	○
		ELECTRICITY		6		○			○	○	○	○	○	○	○
		CONSTRUCTION		7		○			○	○	○	○	○	○	○
		OIL		8		○			○	○	○	○	○	○	○
		TRANSPORT AND COMMUNICATIONS		9		○			○	○	○	○	○	○	○
		SERVICES		10											
	COM-POSITE PUBLIC	FOOD PROCESSING	AGR.C. RENT ACCOUNT	11			○								
			COMPOSITE AGRICULT.	12		·									
PRIM. FACT.	CAPITAL PUBLIC		FOOD PROCESS.	13				○							

TABLE A.1.6

COMPOSITE DOMESTIC INTERMEDIATE INPUTS: PRIVATE ACTIVITIES

					COMPOSITE DOMESTIC INTERMEDIATE INPUTS								
					PRIVATE ACTIVITIES								
					AGRICULTURE	TOTAL INTERMEDIATE	FOOD PROCESSING		TEXTILES	OTHER INDUSTRIES	CONSTRUCTION	TRANSPORT AND COMMUNICATIONS	SERVICES
							COMPOSITE AGRICULTURE	AGRICULTURE RENT ACCOUNT					
					1	2	3	4	5	6	7	8	9
DOM.GOVT. TRA.COM.	AGRICULTURE		SUBSIDIZED VALUE	1				○					
NON-GOVERNMENT COMMODITIES	COMPOSITE DOMESTIC	AGRICULTURE		2	○		○		○	○	○	○	○
		FOOD PROCESSING		3	○	○			○	○	○	○	○
		TEXTILES		4	○	○			○	○	○	○	○
		OTHER INDUSTRIES		5	○	○			○	○	○	○	○
		ELECTRICITY		6	○	○			○	○	○	○	○
		CONSTRUCTION		7	○	○			○	○	○	○	○
		OIL		8	○	○			○	○	○	○	○
		TRANSPORT AND COMMUNICATIONS		9	○	○			○	○	○	○	○
		SERVICES		10	○	○			○	○	○	○	○
COMPOSITE INPUTS	PRIVATE DOMESTIC	FOOD PROCESSING	AGRICULTURE RENT ACCOUNT	11			○						
			COMPOSITE AGRICULTURE	12		○							
PRIMARY FACTORS	PRIVATE CAPITAL		FOOD PROCESSING	13				○					

TABLE A.1./

COMPOSITE IMPORTED INTERMEDIATE INPUTS: PUBLIC ACTIVITIES

			COMPOSITE IMPORTED INTERMEDIATE INPUTS													
			AGRICULTURE			FOOD PROCESSING				PUBLIC ACTIVITIES						
			1 TOTAL INTERMEDIATE	2 COMPOSITE OTHER INDUSTRIES	3 OTHER INDUSTRIES RENT ACCOUNT	4 TOTAL INTERMEDIATE	5 COMPOSITE AGRICULTURE	6 COMPOSITE FOOD PROCESSING	7 AGRICULTURE RENT ACCOUNT	8 FOOD PROCESSING RENT ACCOUNT	9 TEXTILES	10 OTHER INDUSTRIES	11 ELECTRICITY	12 CONSTRUCTION	13 OIL	14 TRANSPORT AND COMMUNICATIONS	15 SERVICES
GOVERNMENT TRADE IMPORTS		1 AGRICULTURE	○				○	○									
		2 FOOD PROCESSING		○													
		3 OTHER INDUSTRIES			○												
NON-GOVERNMENT COMPOSITE IMPORTS		4 AGRICULTURE									○	○					○
		5 FOOD PROCESSING									○	○				○	○
		6 TEXTILES									○	○				○	○
		7 OTHER INDUSTRIES									○	○	○	○	○	○	○
		8 OIL									○	○	○	○	○	○	○
		9 TRANSPORT AND COMMUNICATIONS														○	○
		10 SERVICES														○	○
COMPOSITE INPUTS	PUBLIC IMPORTS	11 TOTAL INTERMEDIATE	○			○											
		12 OTHER INDUST. RENT ACCOUNT		○													
		13 COMPOSITE OTHER INDUST.	○														
PRIM. FACT. CAPITAL		14 PUBLIC AGRICULTURE			○												
COMPOSITE INPUTS	PUBLIC IMPORTS FOOD PROCESSING	15 AGRICULTURE RENT ACCOUNT					○										
		16 FOOD PROCESS. RENT ACCOUNT						○									
		17 COMPOSITE AGRICULTURE				○											
		18 COMPOSITE FOOD PROCESS.				○											
PRIM. FACT. PUBLIC CAPITAL		19 FOOD PROCESSING							○	○							

SUBSIDIZED VALUE

- 138 -

TABLE A.1.8

COMPOSITE IMPORTED INTERMEDIATE INPUTS: PRIVATE ACTIVITIES

				COMPOSITE INPUTS -- IMPORTS — PRIVATE ACTIVITIES														
				AGRICULTURE					FOOD PROCESSING									
				TOTAL INTERMEDIATE	COMPOSITE AGRICULTURE	COMPOSITE OTHER INDUSTRIES	AGRICULTURE RENT ACCOUNT	OTHER INDUSTRIES RENT ACCOUNT	TOTAL INTERMEDIATE	COMPOSITE AGRICULTURE	COMPOSITE FOOD PROCESSING	AGRICULTURE RENT ACCOUNT	FOOD PROCESSING RENT ACCOUNT	TEXTILES	OTHER INDUSTRIES	CONSTRUCTION	TRANSPORT AND COMMUNICATIONS	SERVICES
				1	2	3	4	5	6	7	8	9	10	11	12	13	14	15
GOVERNMENT TRADE IMPORTS		AGRICULTURE	1				○					○						
		FOOD PROCESSING	2	SUBSIDIZED VALUE									○					
		OTHER INDUSTRIES	3					○										
NON-GOVERNMENT IMPORTS	COMPOSITE INPUTS	AGRICULTURE	4		○					○				○	○			○
		FOOD PROCESSING	5							○				○	○		○	○
		TEXTILES	6	○					○					○	○			○
		OTHER INDUSTRIES	7			○			○					○	○	○	○	○
		OIL	8	○					○					○	○	○	○	○
		TRANSPORT AND COMMUNICATIONS	9														○	
		SERVICES	10														○	○
COMPOSITE INPUTS	PRIVATE IMPORTS	AGRICULTURE — AGRICULTURE RENT ACCOUNT	11		○													
		OTHER INDUSTRIES RENT ACCOUNT	12			○												
		AGRICULTURE RENT TRANSFER	13															
		OTHER INDUSTRIES RENT TRANSFER	14															
		COMPOSITE AGRICULTURE	15	○														
		COMPOSITE OTHER INDUSTRIES	16	○														
PRIMARY FACTORS	CAPITAL	PRIVATE AGRICULTURE	17				○	○										
COMPOSITE INPUTS	PRIVATE IMPORTS	FOOD PROCESSING — AGRICULTURE RENT ACCOUNT	18						○									
		FOOD PROCESSING RENT ACCOUNT	19							○								
		AGRICULTURE RENT TRANSFER	20															
		FOOD PROCESSING RENT TRANSFER	21															
		COMPOSITE AGRICULTURE	22						○									
		COMPOSITE FOOD PROCESSING	23						○									
PRIMARY FACTORS	PRIVATE CAPITAL	FOOD PROCESSING	24									○	○					

4.3. Current Institutions Accounts

4.3.1 Household Accounts

Table A.1.9 represents uses of urban and rural household incomes. This table is a disaggregation of column 6 of the intermediate SAM presented in Table A.1.b. Column 1, a revenue account, presents urban household outlays: (i) transfers to rural households; (ii) interest payments on credit to private and public companies; (iii) payments to the social security account; (iv) payments to the government, direct taxes, and (v) transfers to abroad. The remaining entry in the cell 9.1 indicates disposable income of urban households. This disposable income is allocated to the committed expenditure account of urban households at the intersection of column 2 with row 11, and to the private savings pool in row 15. The expenditures of urban households appear in column 3. These are disaggregated into a consumption of government trade subsidized commodities (constrained demand) indicated in entries $T_{16.3}$ to $T_{19.3}$, and a consumption of domestic and imported non-government commodities indicated in $T_{28.3}$ to $T_{42.3}$. Total discretionary expenditure appears in row 13. Discretionary expenditures are disaggregated in column 4, where they are spent in a similar way as committed expenditures: on government trade and non-government trade commodities (row 16 to 19, and 28 to 42).

Accounts in columns 5 to 8 represent disaggregation of the constrained urban household demand. The diagonal entries in these columns represent the values that households are willing to pay for government trade commodities. The entries in row 9, on the other hand, represent income transfer occuring to urban households due to rationing and subsidization of government trade commodities.

The other half of Table A.1.9 describes expenditures of rural households income. Conceptually, it is identical to the left part of the table. The only difference is in the magnitude of actual numbers which reflect the differences in expenditures patterns between urban and rural households.

4.3.2 Companies Accounts

Table A.1.10 describes expenditures accounts of private and public companies. The first entry in column 1 represents dividends paid by private companies to urban households. The second entry indicates transfers from private companies to other public companies. The entry in row six represents savings of private companies, while direct taxes are paid in row 12. These entries are the same as the ones that appear in column 7 of the intermediate SAM in Table A.1.b.

The second column of Table A.1.10 describes expenditues of EGPC on: dividends paid to urban households, transfers to other public companies, savings, interest payments to abroad, transfered profits to the government, and payments of taxes.

The third column in this table describes expenditures of other public companies. In principle, all of the entries in this column are the same as in the first two columns, except that interest payments to abroad are channeled

TABLE A.1.9
HOUSEHOLD ACCOUNTS

				URBAN HOUSEHOLDS								RURAL HOUSEHOLDS							
				CURRENT ACCOUNT INSTITUTIONS				CONSTRAINED DEMAND				CURRENT ACCOUNT INSTITUTIONS				CONSTRAINED DEMAND			
										GOVERNMENT COMMODITIES								GOVERNMENT COMMODITIES	
				REVENUE ACCOUNT	DISPOSABLE INCOME	COMMITTED EXPENDITURES	DISCRETIONARY EXPENDITURES	AGRICULTURE	FOOD PROCESSING	OTHER INDUSTRIES	SERVICES	REVENUE ACCOUNT	DISPOSABLE INCOME	COMMITTED EXPENDITURES	DISCRETIONARY EXPENDITURES	AGRICULTURE	FOOD PROCESSING	OTHER INDUSTRIES	SERVICES
				1	2	3	4	5	6	7	8	9	10	11	12	13	14	15	16
HOUSEHOLDS	URBAN	REVENUE ACCOUNT	1																
	RURAL		2	○															
CURRENT ACCOUNT INSTITUTIONS	PRIVATE COMPANIES		3	○								○							
	OTHER PUBLIC COMPANIES		4	○								○							
	SOCIAL SECURITY		5	○								○							
	GOVERNMENT REVENUE		6	○								○							
	DIRECT TAXES		7	○								○							
	R.O.W.2 COMMERCIAL BANKS		8	○								○							
HOUSEHOLDS	URBAN	DISPOSABLE INCOME	9	○				○○○○											
	RURAL		10									○				○○○○			
	URBAN	COMMITTED EXPENDITURES	11		○														
	RURAL		12										○						
	URBAN	DISCRETIONARY EXPENDITURES	13				○												
	RURAL		14												○				
	PRIVATE SAVINGS POOL		15		○								○						
CONSTRAINED DEMAND GOVERNMENT COMMODITIES HOUSEHOLDS	URBAN	AGRICULTURE	16			○○○													
		FOOD PROCESSING	17			○○○													
		OTHER INDUSTRIES	18			○○○													
		SERVICES	19			○○○													
	RURAL	AGRICULTURE	20											○○○					
		FOOD PROCESSING	21											○○○					
		OTHER INDUSTRIES	22											○○○					
		SERVICES	23											○○○					
GOVERNMENT COMPOSITE COMMODITIES HOUSEHOLD DEMAND		AGRICULTURE	24					○								○			
		FOOD PROCESSING	25						○								○		
		OTHER INDUSTRIES	26							○								○	
		SERVICES	27								○								○
NON-GOVERNMENT COMMODITIES COMPOSITE DOMESTIC		AGRICULTURE	28			○○								○○					
		FOOD PROCESSING	29			○○								○○					
		TEXTILES	30			○○								○○					
		OTHER INDUSTRIES	31			○○								○○					
		ELECTRICITY	32			○○								○○					
		OIL	33			○○								○○					
		TRANSPORT AND COMMUNICATIONS	34			○○								○○					
		SERVICES	35			○○								○○					
NON-GOVERNMENT IMPORTS COMPOSITE IMPORTED		AGRICULTURE	36			○○								○○					
		FOOD PROCESSING	37			○○								○○					
		TEXTILES	38			○○								○○					
		OTHER INDUSTRIES	39			○○								○○					
		OIL	40			○○								○○					
		TRANSPORT AND COMMUNICATIONS	41			○○								○○					
		SERVICES	42			○○								○○					

TABLE A.1.10
COMPANIES CCOUNTS

				CURRENT ACCOUNT INSTITUTIONS		
				PRIVATE COMPANIES	PUBLIC COMPANIES	
					EGPC	OTHERS
				1	2	3
CURRENT ACCOUNT INSTITUTIONS	URBAN HOUSEHOLDS	REVENUE ACCOUNT	1	◯	◯	◯
	RURAL HOUSEHOLDS	REVENUE ACCOUNT	2			◯
	OTHER PUBLIC COMPANIES		3	◯	◯	
	PRIVATE COMPANIES		4			◯
	PUBLIC COMPANIES EGPC		5			◯
CAPITAL ACCOUNT	PRIVATE SAVINGS POOL		6	◯		
	PUBLIC SAVINGS POOL		7		◯	◯
CURRENT ACCOUNT INSTITUTIONS	R.O.W. 1 CENTRAL BANK		8		◯	◯
	R.O.W. 2 COMMERCIAL BANK		9			◯
	GOVERNMENT REVENUE		10		◯	◯
	SOCIAL SECURITY		11		◯	◯
	DIRECT TAXES		12	◯	◯	◯

through both, the Central Bank and the commercial banks pool (rows 8 and 9), and that the entry in row 11 indicates payments to the social security account.

4.3.3 Government, Social Security and the Tax Accounts

Table A.1.11 describes the government accounts. Government income is derived from row 1, i.e. the tax accounts and government trade. It should be noted that entries for subsidies (column 11) and government trade (column 8) are negative, since both of these accounts are financed by the government. This means that the sum of the row 1 gives the net revenue of the government.

Government expenditures appear in columns 1 to 7 in Table A.1.11. Government incomes are at first allocated to the government committed expenditures accounts for education, health, and other government at the intersection of rows 2 to 4 with column 1. Government transfers to urban and rural households and companies are indicated in rows 5 to 8. Social security contributions and savings appear in rows 9 and 10 and interest payments, including transfers abroad in row 35.

Committed and discretionary expenditures of the three government sectors: education, health, and others are indicated in columns 2 to 7 in Table A.1.11. Government expenditure on labor is indicated in row 11, while government consumption of domestic and imported government trade (subsidized) commodities appears in rows 12 to 15. Discretionary expenditures appear on diagonal entries in rows 16 to 18. These are then disaggregated as government consumption of domestic and imported non-government composite commodities in columns 5 to 7 and rows 19 to 34.

Social security revenues appear in row 9 and expenditures in column 9. Expenditures consist of social security payments to urban and rural households, payments to other companies, surplus of the social security account ($T_{10.9}$), and social security payments to the government labor (row 11).

4.3.4 The Rest of the World Accounts

Table A.1.12 describes the accounting structure of the rest of the world account. As indicated in section 3.2 above, there are three different foreign exchange rates currently operating in Egypt. The detailed accounting relationships between these three foreign exchange pools, the transfer account, and the rest of the SAM accounts are presented in columns 1 to 4 in Table A.1.12.

Column 1 of the rest of the world account represents outflows of foreign exchange from the Central Bank pool. The sum of this column is measured at the Central Bank foreign exchange rate. Individual entries in column 1, from top to the bottom, are indicated at the intersections with non empty row cells. The first entry in row 2 indicates receipts by primary factors, i.e., earnings of the national capital or interest payments of the rest of the world to Egypt. Transfers from abroad to the government are indicated in row 6. Net public sector borrowing from abroad (at the Central Bank exchange rate) appears in row 7. Entries in rows 9 and 10 indicate payment of the rest of the world for government and non-government trade

TABLE A.1.11
GOVERNMENT, SOCIAL SECURITY, AND THE TAX ACCOUNTS

						CURRENT ACCOUNT INSTITUTIONS												
						GOVERNMENT								TAXES				
						COMMITTED			DISCRETIONARY									
						GOVERNMENT REVENUE	EDUCATION	HEALTH	OTHER	EDUCATION	HEALTH	OTHER	GOVERNMENT TRADE	SOCIAL SECURITY	INDIRECT	SUBSIDY	DIRECT	IMPORT TARIFFS	EXPORT TAXES
						1	2	3	4	5	6	7	8	9	10	11	12	13	14
CURRENT ACCOUNT INSTITUTIONS	GOVERNMENT		REVENUE		1								○		○	○	○	○	○
		COMMITTED	EDUCATION		2	○													
			HEALTH		3	○													
			OTHER		4	○													
	HOUSEHOLDS	URBAN	REVENUE		5	○								○					
		RURAL	ACCOUNT		6	○								○					
	PUBLIC COMPANIES		EGPC		7	○													
			OTHERS		8	○								○					
	SOCIAL SECURITY				9	○													
CAPITAL ACCOUNT	PUBLIC SAVINGS POOL				10	○								○					
TOTAL LABOR	GOVERNMENT				11		○	○	○				○						
DOMESTIC GOVT. TRADE COMMODITIES		SUBSIDIZED VALUE	AGRICULTURE		12		○	○	○										
			FOOD PROCESSING		13		○	○	○										
GOVERNMENT TRADE IMPORTS			AGRICULTURE		14		○	○	○										
			FOOD PROCESSING		15		○	○	○										
CURRENT ACCOUNT INSTITUTIONS	GOVERNMENT DISCRETIONARY		EDUCATION		16	○													
			HEALTH		17		○												
			OTHER		18		○												
NON-GOVERNMENT COMMODITIES COMPOSITE DOMESTIC			AGRICULTURE		19					○	○	○							
			FOOD PROCESSING		20					○	○	○							
			TEXTILES		21					○	○	○							
			OTHER INDUSTRIES		22					○	○	○							
			ELECTRICITY		23					○	○	○							
			CONSTRUCTION		24					○	○	○							
			OIL		25					○	○	○							
			TRANSPORT AND COMMUNICATIONS		26					○	○	○							
			SERVICES		27					○	○	○							
NON-GOVERNMENT IMPORTS COMPOSITE IMPORTS			AGRICULTURE		28					○	○	○							
			FOOD PROCESSING		29					○	○	○							
			TEXTILES		30					○	○	○							
			OTHER INDUSTRIES		31					○	○	○							
			OIL		32					○	○	○							
			TRANSPORT AND COMMUNICATIONS		33					○	○	○							
			SERVICES		34					○	○	○							
CURRENT ACCOUNT INSTITUTIONS	REST OF THE WORLD 1	CENTRAL BANK			35	○													

TABLE A.1.12
REST OF THE WORLD ACCOUNTS

				CURRENT ACCOUNT INSTITUTIONS			
				REST OF WORLD 1	TRANSFER	REST OF WORLD 2	REST OF WORLD 3
				CENTRAL BANK	REST OF WORLD 1 AND 2	COMMERCIAL BANKS	OWN EXCHANGE
				1	2	3	4
PRIMARY FACTORS	LABOR ABROAD		1			○	○
	NATIONAL CAPITAL		2	○		○	
CURRENT ACCOUNT INSTITUTIONS	HOUSEHOLD REVENUE ACCOUNT	URBAN	3			○	
		RURAL	4			○	
	OTHER PUBLIC COMPANIES		5			○	
	GOVERNMENT REVENUE		6	○			
CAPITAL ACCOUNT	SAVINGS POOL	PUBLIC	7	○		○	
		PRIVATE	8				○
GOVT. TRADE EX.	PRIVATE ACTIV.	AGRICULTURE	9	○			
NON-GOVERNMENT COMMODITIES	COMPOSITE EXPORTS	AGRICULTURE	10	○			
		FOOD PROCESSING	11				○
		TEXTILES	12				○
		OTHER INDUSTRIES	13				○
		OIL	14	○			
		TRANSPORT AND COMM.S. — SUEZ CANAL	15	○			
		TRANSPORT AND COMM.S. — OTHER	16			○	
		SERVICES	17			○	
CURRENT ACCOUNT INSTITUTIONS	ROW 1 POOL	CENTRAL BANK	18		○		
		IMPORT PREMIA	19		○		
	TRANSFER	ROW 1 AND ROW 2	20			○	

exports of agricultural goods, and the remaining two cells in column 1 represent payments of the rest of the world for non-government exports of oil and Suez Canal services.

The second column is a transfer account between the Central Bank pool and the commercial banks pool. The intersection of row 18 with column 2 indicates transfers of foreign exchange from the Central Bank pool to the commercial banks pool at the Central Bank exchange rate. Because the sum of this column is valued at the commercial banks rate, the next entry in row 19 and column 2 represents the difference of valuation of the foreign exchange.

The third column represents the outflows of foreign exchange from the commercial banks pool. The sum of this column is valued at the commercial banks foreign exchange rate. First five entries in the column indicate remittances, capital earnings, and transfers from abroad to urban and rural households and other public companies. Next entry in cell 7.3 indicates net public sector borrowing from abroad at the commercial banks exchange rate. Entries in rows 11 to 17 represent payments of the commercial banks pool for non-government exports, and $T_{20.3}$ indicates payments of the commercial banks pool (in Egyptian pounds) for foreign exchange received from the Central Bank pool through the transfer account in row 20.

Column four indicates outflows of foreign exchange from the own exchange budget. The sum of this column is measured at the own exchange rate. The first entry in column four indicates remittances and the second entry represents net private sector borrowing from abroad valued at the own exchange rate.

4.4. Capital Accounts

Table A.1.13 describes expenditures on the Capital Accounts. Column 1 of this table indicates allocation of private savings to the public savings pool and private investment account (rows 1 and 2). Column 2 shows private sector investment by sector of destination. Column 3 shows sources of investment for the government and state enterprises. The government investment income received in row 3 from the public savings pool is spent in column 4, where entries in rows 23 to 40 represent fixed capital formation and change in stocks by the government. Column 5 represents investment by sector of destination by public enterprises.

Public and private activities investment is captured in columns 6 to 21. Because of a lack of data, public and private investments are aggregated into two accounts through the capital goods account in rows 21 and 22. Columns 22 and 23 of the capital goods account then show aggregate private and public activities investment by destination.

4.5 Activities

4.5.1 Public Activities

Table A.1.14 describes expenditures (cost structure) by public activities. In this table, each activity column is measured at three different prices: factor cost, producers cost, and users cost. For example, in column 1 public agriculture pays for services of primary factors of

TABLE A.1.13
CAPITAL ACCOUNTS

			CAPITAL ACCOUNT																						
					INVEST-MENT		PUBLIC ACTIVITIES								PRIVATE ACTIVITIES								CAPITAL GOODS	
			PRIVATE SAVINGS POOL	PRIVATE INVESTMENT	PUBLIC SAVINGS POOL	GOVERNMENT	STATE ENTERPRISES	AGRICULTURE	FOOD PROCESSING	TEXTILES	OTHER INDUSTRIES	ELECTRICITY	CONSTRUCTION	OIL	TRANSPORT & COMM.	SERVICES	AGRICULTURE	FOOD PROCESSING	TEXTILES	OTHER INDUSTRIES	CONSTRUCTION	TRANSPORT & COMM.	SERVICES	PRIVATE SECTOR	PUBLIC SECTOR
			1	2	3	4	5	6	7	8	9	10	11	12	13	14	15	16	17	18	19	20	21	22	23
	PUBLIC SAVINGS POOL	1	○																						
	PRIVATE INVESTMENT	2	○																						
	GOVERNMENT INVESTMENT	3			○																				
	STATE ENTERPRISE INVEST	4			○																				
CAPITAL ACCOUNT / PUBLIC ACTIVITIES	AGRICULTURE	5					○																		
	FOOD PROCESSING	6					○																		
	TEXTILES	7					○																		
	OTHER INDUSTRIES	8					○																		
	ELECTRICITY	9					○																		
	CONSTRUCTION	10					○																		
	OIL	11					○																		
	TRANSPORT & COMM.	12					○																		
	SERVICES	13					○																		
CAPITAL ACCOUNT / PRIVATE ACTIVITIES	AGRICULTURE	14		○																					
	FOOD PROCESSING	15		○																					
	TEXTILES	16		○																					
	OTHER INDUSTRIES	17		○																					
	CONSTRUCTION	18		○																					
	TRANSPORT & COMM.	19		○																					
	SERVICES	20		○																					
CAPITAL GOODS	PRIVATE SECTOR	21															○	○	○	○	○	○	○		
	PUBLIC SECTOR	22						○	○	○	○	○	○	○	○	○									
DOM.GOVT. TR.CO. VALUE	AGRICULTURE	23				○																			
GOVT. TRADE IMP. SUBS.	AGRICULTURE	24				○																			
	FOOD PROCESSING	25				○																			
NON-GOVERNMENT COMPOSITE DOMESTIC COMMODITIES	AGRICULTURE	26																							○
	FOOD PROCESSING	27																							○
	TEXTILES	28																							○
	OTHER INDUSTRIES	29				○																		○	○
	ELECTRICITY	30																							○
	CONSTRUCTION	31				○																		○	○
	OIL	32																						○	○
	TRANSPORT & COMM.	33				○																		○	○
	SERVICES	34				○																		○	○
NON-GOVERNMENT COMPOSITE IMPORTS	AGRICULTURE	35																						○	○
	FOOD PROCESSING	36																						○	○
	TEXTILES	37																							○
	OTHER INDUSTRIES	38				○																		○	○
	OIL	39																							
	TRANSPORT & COMM.	40				○																		○	○

TABLE A.1.14
PUBLIC ACTIVITIES

			PUBLIC ACTIVITIES																							
			AGRICULTURE			FOOD PROCESSING			TEXTILES			OTHER INDUSTRIES			ELECTRICITY			CONSTRUCTION			OIL			TRANSPORT & COMMUNICATION			SERVICES		
			FACTOR COST	PRODUCERS COST	USER COST	FACTOR COST	PRODUCERS COST	USER COST	FACTOR COST	PRODUCERS COST	USER COST	FACTOR COST	PRODUCERS COST	USER COST	FACTOR COST	PRODUCERS COST	USER COST	FACTOR COST	PRODUCERS COST	USER COST	FACTOR COST	USER COST	FACTOR COST	PRODUCERS COST	USER COST	FACTOR COST	PRODUCERS COST	USER COST
			1	2	3	4	5	6	7	8	9	10	11	12	13	14	15	16	17	18	19	20	21	22	23	24	25	26
AGRICULTURE	LAND	1	○																									
	CAPITAL	2	○																									
	LABOR	3	○																									
	DOMESTIC COMP. INPUTS	4	○																									
	IMPORTED COMP. INPUTS	5	○																									
	FACTOR COST	6		○																								
	PRODUCERS COST	7			○																							
FOOD PROCESSING	CAPITAL	8				○			○																			
	LABOR	9				○																						
	DOMESTIC COMP. INPUTS	10				○																						
	IMPORTED COMP. INPUTS	11				○																						
	FACTOR COST	12					○																					
	PRODUCERS COST	13						○																				
TEXTILES	CAPITAL	14							○			○																
	LABOR	15							○																			
	DOMESTIC COMP. INPUTS	16							○																			
	IMPORTED COMP. INPUTS	17							○																			
	FACTOR COST	18								○																		
	PRODUCERS COST	19									○																	
OTHER INDUSTRIES	CAPITAL	20										○			○													
	LABOR	21										○																
	DOMESTIC COMP. INPUTS	22										○																
	IMPORTED COMP. INPUTS	23										○																
	FACTOR COST	24											○															
	PRODUCERS COST	25												○														
ELECTRICITY	CAPITAL	26													○			○										
	LABOR	27													○													
	DOMESTIC COMP. INPUTS	28													○													
	IMPORTED COMP. INPUTS	29													○													
	FACTOR COST	30														○												
	PRODUCERS COST	31															○											
CONSTRUCTION	CAPITAL	32																○			○							
	LABOR	33																○										
	DOMESTIC COMP. INPUTS	34																○										
	IMPORTED COMP. INPUTS	35																○										
	FACTOR COST	36																	○									
	PRODUCERS COST	37																		○								
OIL	CAPITAL	38																			○							
	LABOR	39																			○							
	DOMESTIC COMP. INPUTS	40																			○							
	IMPORTED COMP. INPUTS	41																			○							
	FACTOR COST	42																				○						
	PRODUCERS COST	43																						○				
TRANSPORT AND COMMUNICATIONS	CAPITAL	44																					○					
	LABOR	45																					○					
	DOMESTIC COMP. INPUTS	46																					○					
	IMPORTED COMP. INPUTS	47																					○					
	FACTOR COST	48																						○				
	PRODUCERS COST	49																									○	○
SERVICES	CAPITAL	50																								○		
	LABOR	51																								○		
	DOMESTIC COMP. INPUTS	52																								○		
	IMPORTED COMP. INPUTS	53																								○		
	FACTOR COST	54																									○	
	PRODUCERS COST	55																										
CURRENT ACCOUNT INSTITUTIONS	INDIRECT TAXES	56	○			○			○			○			○			○			○		○			○		

production in rows 1 to 3 (land, capital and labor) and buys domestic and imported intermediate inputs from row accounts 4 and 5. The total of this column represents factor cost of public agriculture. In column 2 public agriculture buys primary factors (including intermediate inputs) from row 6 and pays taxes (fees and licences) to the tax account in row 56. The sum of this column indicates producers cost of public agriculture. The total of column 3 (users cost) is in this case equal to column 2, because there is no difference between producers cost and users cost for agriculture.

The accounting structure of other public activities in Table A.1.14 is the same as for agriculture with exception that there is a difference between producers and users prices. This difference is due to fixed prices determined by the government for public production. Fixed prices are reflected in Table A.1.14, for example, for food processing activity at the intersection of column 6 with row 8, where $T_{8.6}$ indicates a negative rent which occurs because of fixed prices. All other public activities are treated in the same way as food processing activity with the exception of oil (column 19 and 20), where the producers cost is equal to the users cost.

4.5.2. Private Activities

Table A.1.15 describes the cost structure of private activities. This table is conceptually the same as Table A.1.14. However, because prices for private activities are not controlled by the government, there is no difference between the producers cost and the users cost of private activities. For example, private agriculture pays for services of primary factors of production and for intermediate inputs in column 1, thus giving factor cost of production. In column two private activity buys primary inputs from row 6 and pays taxes in row 37. This column determines producers cost of production. All other private activities are treated in the same way.

4.6 Commodities

4.6.1. Government Trade Commodities

Table A.1.16 describes government trade commodities. For example, columns 1 to 4 represent a demand for subsidized government trade domestic and imported commodities for agriculture, food processing, other industries, and services. These commodities are sold in Table A.1.9 (presented above) from diagonal entries in rows 24 to 27 to urban and rural household (constrained demand) account. In essence, columns 1 to 4 in Table A.1.16 give the composition between domestics and imports of the government trade commodities consumed by households.

The second set of accounts in Table A.1.16, columns 5 to 11, shows the composition of domestic government trade commodities. For example, government trade agriculture account (column 5) buys agricultural commodities from private activities in row 9. Because agricultural commodity is bought by the government trade at fixed prices, there is an additional entry at the intersection of column 5 with row 24. This entry represents a negative rent which accrues to the capital employed in private agriculture. The sum of column 5 is sold to the column 6 from row 12 where trade margins (services) are entered at the intersection of column of 6 with row 11. Column 6 thus represents the cost of private agricultural commodity to the government.

TABLE A.1.15
PRIVATE ACTIVITIES

				PRIVATE ACTIVITIES												
				AGRI-CULTURE		FOOD PROCESS		TEX-TILES		OTHER INDUST.		CONS TRUCT.		TRANSP. & COMM.		SER-VICES
				FACTOR COST	PRODUCERS COST	FACTOR COST	PRODUCERS COST	FACTOR COST	PRODUCERS COST	FACTOR COST	PRODUCERS COST	FACTOR COST	PRODUCERS COST	FACTOR COST	PRODUCERS COST	FACTOR COST	PRODUCERS COST
				1	2	3	4	5	6	7	8	9	10	11	12	13	14
P R I V A T E A C T I V I T I E S	AGRICULTURE	LAND	1	○													
		CAPITAL	2	○													
		LABOR	3	○													
		DOMESTIC COMP. INPUTS	4	○													
		IMPORTED COMP. INPUTS	5	○													
		FACTOR COST	6		○												
	FOOD PROCESSING	CAPITAL	7			○											
		LABOR	8			○											
		DOMESTIC COMP. INPUTS	9			○											
		IMPORTED COMP. INPUTS	10			○											
		FACTOR COST	11				○										
	TEXTILES	CAPITAL	12					○									
		LABOR	13					○									
		DOMESTIC COMP. INPUTS	14					○									
		IMPORTED COMP. INPUTS	15					○									
		FACTOR COST	16						○								
	OTHER INDUSTRIES	CAPITAL	17							○							
		LABOR	18							○							
		DOMESTIC COMP. INPUTS	19							○							
		IMPORTED COMP. INPUTS	20							○							
		FACTOR COST	21								○						
	CONSTRUCTION	CAPITAL	22									○					
		LABOR	23									○					
		DOMESTIC COMP. INPUTS	24									○					
		IMPORTED COMP. INPUTS	25									○					
		FACTOR COST	26										○				
	TRANSPORT & COMMUNICATIONS	CAPITAL	27											○			
		LABOR	28											○			
		DOMESTIC COMP. INPUTS	29											○			
		IMPORTED COMP. INPUTS	30											○			
		FACTOR COST	31												○		
	SERVICES	CAPITAL	32													○	
		LABOR	33													○	
		DOMESTIC COMP. INPUTS	34													○	
		IMPORTED COMP. INPUTS	35													○	
		FACTOR COST	36														○
CURRENT ACCOUNT INSTITUTIONS		INDIRECT TAXES	37	○		○		○		○		○		○		○	

TABLE A.1.16

GOVERNMENT TRADE COMMODITIES (DOMESTIC, IMPORTS AND EXPORTS)

The subsidized value of this commodity then appears in column 7. The amount of subsidy is indicated at the intersection of row 16 with column 7 as a negativeentry received by the government trade account. Columns 8 to 10 are treated in a similar way with an exception that private food processing commodities are not traded at fixed prices. Column 11, government trade services has only one positive entry which indicates surplus of government trade on this account.

Government trade imports, columns 12 to 17, are treated in a similar way as domestic government trade commodities. For example, government trade imports of agriculture are bought in column 12 from the rest of the world (Central Bank pool) in row 17. Import tariffs appear in row 18. The sum of column 12 gives landed value of agricultural imports. Subsidized value of these imports is indicated in column 13, where subsidies appear in the government trade account (row 16) and landed value of agricultural imports in row 19. The remaining import columns from 14 to 17 are treated in exactly the same way as columns 12 and 13.

The last account of this table, government trade exports appear in columns 18 to 20. Column 18 determines the fixed price of agricultural commodity to private producer in the same way as column 5 of this table. Trade margins are entered at the intersection of column 19 with row 11, and export taxes are indicated at the intersection of row 25 with column 20. Column 20 thus represents the price at which private agricultural commodities are sold by the government trade to the rest of the world.

4.6.2. Non-government Commodities

Tables A.1.17 and A.1.18 represent the accounting structure of non-government commodities. In Table A.1.17 commodities are distinguished by domestic public and private and by public and private exports. Composite domestic commodities and composite exports are a combination of private and public commodities presented in Table A.1.18.

Domestic public non-government commodities are presented in columns 1 to 9 in Table A.1.17. Each column has two entries, for example, agriculture in column 1 buys output from public activity agriculture at users cost from row 1, and trade margins from services in row 9. The sums of columns 1 to 9 are sold from rows 17 to 25 to the domestic public rent account columns. Columns 10 to 18 of the latter account have two entries. The diagonal entries represent supply of domestic public non-government commodities at users cost plus trade margins, while entries in row 26, columns 10 to 18 indicate rents which accrue on the capital account because of rationing.[1]

Domestic private non-government commodities (columns 19 to 25) are treated in a similar way with exception of the rent account. The diagonal entries in these columns represent output bought from activities (at producers cost) and entries in row 16 indicate trade margins. Public and private exports, columns 26 to 38, are treated in exactly the same way as domestic private commodities.

---

[1] Demand is available at fixed price but in rationed amount.

TABLE A.1.17
NON-GOVERNMENT COMMODITIES (DOMESTIC AND EXPORTS)

TABLE A.1.18
NON-GOVERNMENT COMPOSITE COMMODITIES (DOMESTIC AND EXPORTS)

			NON-GOVERNMENT COMMODITIES														
			COMPOSITE DOMESTIC								COMPOSITE EXPORTS						
			AGRICULTURE	FOOD PROCESSING	TEXTILES	OTHER INDUSTRIES	ELECTRICITY	CONSTRUCTION	OIL	TRANSPORT & COMMUNICATIONS	SERVICES	AGRICULTURE	FOOD PROCESSING	TEXTILES	OTHER INDUSTRIES	OIL	TRANSPORT AND COMMUNICATIONS	SERVICES
			1	2	3	4	5	6	7	8	9	10	11	12	13	14	15	16
CURRENT ACCOUNT INSTITUTIONS	INDIRECT TAXES	1	O	O	O	O	O	O	O	O	O							
	SUBSIDIES	2			O	O		O	O	O								
PRIMARY FACTORS	PUBLIC CAPITAL	3														O		
NON-GOVERNMENT / DOMESTIC PUBLIC RENT	AGRICULTURE	4	O															
	FOOD PROCESSING	5		O														
	TEXTILES	6			O													
	OTHER INDUSTRIES	7				O												
	ELECTRICITY	8					O											
	CONSTRUCTION	9						O										
	OIL	10							O									
	TRANSPORT AND COMMUNICATIONS	11								O								
	SERVICES	12									O							
DOMESTIC PRIVATE	AGRICULTURE	13	O															
	FOOD PROCESSING	14		O														
	TEXTILES	15			O													
	OTHER INDUSTRIES	16				O												
	CONSTRUCTION	17						O										
	TRANSPORT AND COMMUNICATIONS	18								O								
	SERVICES	19									O							
PUBLIC EXPORTS	AGRICULTURE	20										O						
	FOOD PROCESSING	21											O					
	TEXTILES	22												O				
	OTHER INDUSTRIES	23													O			
	OIL	24														O		
	TRANSPORT AND COMMUNICATIONS	25															O	
	SERVICES	26																O
PRIVATE EXPORTS	AGRICULTURE	27										O						
	FOOD PROCESSING	28											O					
	TEXTILES	29												O				
	OTHER INDUSTRIES	30													O			
	TRANSPORT AND COMMUNICATIONS	31															O	
	SERVICES	32																O

The accounts presented in Table A.1.18 columns 1 to 9, represent composite domestic non-government commodities. These are valued at market prices which are derived by summing up values of domestic public and private commodities (including trade margins) and indirect taxes and subsidies. For example, in column 3 (textiles), indirect taxes appear in row 1, subsides in row 2, the value of domestic public textiles in row 6, and the value of domestic private textiles in row 15.

Composite exports are presented in Table A.1.18, columns 10 to 16. These are composed of public and private exports shown on diagonals in rows 20 to 32. Export taxes for oil appear in column 14. These are treated as a receipt of the capital account in row 3.

4.6.3. Non-government Imports

Table A.1.19 represents the accounting for non-government imports. Aggregate demand for imports is represented by the composite imports account, columns 31 to 37. Composite imports are a sum of imports traded through the three channels: Central Bank, commercial banks (imports premia), and own exchange pool, plus import tariffs. Import tariffs are indicated at the intersection of row 9 with columns 31 to 37. Central Bank imports appear on the diagonals in rows 10 to 18; the imports premia accounts (commercial banks) are presented in rows 28 to 36; and own exchange imports appear in rows 37 to 45. The sum of columns 31 to 37 gives aggregate demand for imports at market prices.

Individual import channels are presented in columns 3 to 30. The Central Bank imports are indicated in columns 3 to 9, where values in row 6 indicate the actual value of the Central Bank imports and entries in row 5 indicate premia which occurs because of the rationing of these imports.

Imports financed by the discretionary foreign exchange budget appear in columns 10 to 16. Financial sources for these imports are indicated in column 1 of Table A.1.18. The next import channel (commercial banks) is presented in columns 17 to 23 where diagonal entries represent actual price of imports and entries in row 5 indicate import premia which is due to rationing. This import premia is allocated as receipts to urban and rural households and public companies in column 2 in this table. Finally, own exchange imports are presented in columns 24 to 30. As noted above, imports from the three channels are sold from row accounts to the composite imports accounts in order to derive the aggregate demand for imports at market prices.

The above description gives a complete documentation of the accounting framework of the MISR2 model. The accounts also represent the economics of the model. We now turn in the next part of this chapter to the TV formulation and specifications of the model.

C. The MISR2 TV

The previous part of the chapter presented the accounts of the MISR2 models organized in a Social Accounting Matrix (SAM). As noted above each cell of the SAM is a value of a transaction which is the outcome of agents behaviors and constraints ensuring the consistency of these behaviors. Formulating a model in TV amounts to: (i) giving an analytical function

TABLE A.1.19
NON-GOVERNMENT IMPORTS

explaining the behavior of each cell in the SAM and (ii) telling the constraints imposed on each account which ensure the balancing of all accounts and define equilibrium. In the following, a brief technical presentation of the TV concepts and TV specifications is provided.

Let $T = \|t_{ij}\|$ be a SAM with elements t_{ij}, where i is a row and j a column index, i,j=1,2,...,n where n is the size of the SAM. Let $y = \|y_i\|$ be a column vector of total outlays:

$$y_i = \sum_{i=1}^{n} t_{ij} \quad , \quad j = 1, 2, \ldots, n.$$

to a subset of the accounts $m \leq n$ one can associate a price, hence a column vector $p = \|p_k\|$, $k = 1, 2, \ldots, m \leq n$. by definition, for all accounts with prices: $y_j = p_j q_j$; a vector $q_k = \|q_k\|$, $k = 1, 2, \ldots, m \leq n$. of quantities can thus be defined. Finally let $\mu = \|\mu_s\|$, $s = 1, 2, \ldots, S$ be a vector of parameters. Each element t_{ij} can be associated with a function of the general form[1]

$$t_{ij} = t_{ij}(y, p, \mu) . \tag{1}$$

Providing analytical forms for the functions in (1) and telling the nature of each variable y_i, p_k, q_k -- endogenous, exogenous [2] will define an economy-wide equilibrium model.

1. The TV Specifications used in MISR2

Table 1 gives the list of specifications of the $t_{ij}(y, p, \mu)$ functions used in the MISR2 models. Specification 1 tells that the entry t_{ij} in column j is equal to the column sum y_j. In other terms all the outlays of account j are going to account i. Specification 2 simply states that the entry t_{ij} has no specific functional forms. It is a "floating" variable which will be determined by the over-all systems constraints. Typically if government revenue is a function of income and government expenditure a policy variable then government savings will be a residual and given by specification

---

[1] For more detailed description of the TV concepts see: Drud, Grais and Pyatt (1983).

[2] Or undefined. Some accounts recording only transfers may not have a price associated with them. One can only define the total outlays and receipts of those accounts as values. In this case the price and quantity are undefined.

2. Specification 3 indicates that the value of the cell is exogenous. The function $f_{ij}(\theta)$ of time θ indicates an exogenous shift factor. When a cell in a column behaves like a percentage share of the other elements in the column, specification 4 will be used. This will be the case of an advalorem tax or tarif, a fixed markup or a fixed wage-differential. Sometimes elements in a column are specified as value shares of the column total with the possibility of changing through time these value shares. For example, the allocation of disposable income between savings and consumption could be modeled in this way. Similarly the allocation of investment funds over various investors could be represented with within-period constant value shares. In this case specification 5 will be used. If the entry t_{ij} is a payment for a good in completely inelastic demand, it will behave according to specification 6: whatever the price p_i, the demand by the agent represented in column j for the commodity supplied by account i will be $t^o_{ij} f_{ij}(\theta)$ where $f_{ij}(\theta)$ is a shift factor. Specification 6 will typically be used to represent a committed demand in an expenditure system. 1/ Specification 7 indicates that t_{ij} is an expenditure associated with a downward sloping export demand function. In 7, p_j is the exchange rate, p_i the supply price of home goods, p_j is the exchange rate, p_i the supply price of home goods, π^x_i the world price in foreign currency of substitutable goods on the world market, η_{ij} the price-elasticity of demand and $f_{ij}(\theta)$ a shift factor. When the elasticity η_{ij} takes the value zero and the world demand is completely inelastic, specification 7 reduces to specification 8. If on the contrary the world demand for exports is perfectly elastic, exports are determined residually and the supply price of exports has to equal the world price times the exchange rate. This situation will be represented by specification 9. Let y_i be the value of imports including the import tariff, the pre-tariff value of imports will be given by specification 10 where p_i is the exchange rate and π^m_j the price of imports in foreign currency. Assume that the quantity associated with a column is a Constant Elasticity of Substitution (CES) aggregation of the quantities associated with each cell appearing in the column: a CES production function. Under cost minimization, the payment in each cell of the column will behave like specification 11. It is an expenditure function on the inputs entering a CES production function, under

---

1/ See Deaton and Muellbauer (1980) for systems of consumption functions.

cost minimization.$\underline{1/}$ Specification 19 is the same as specification 11 but allowing for technical change through the shift function $f_{ij}(\theta)$. Specification 12 is a special case of 11 where $\sigma_j = 0$: it is an expenditure function associated with an input entering a Leontief aggregation. Specification 15 is identical to 12 except that the share θ_{ij} is not derived from the base-year SAM and can be shifted exogenously between periods. Another rule of allocation of a column sum over the elements entering the column is given by specification 16. The shares allocating y_j are logistic functions of the real value of the total outlay (y_j/p_j). In the base year $y_j/p_j = y_j^0$ and $t_{ij} = t_{ij}^0$.

When y_j changes period changes, the shares will grow or decline logistically according to whether $\rho_{ij} > 0$ or $\rho_{ij} < 0$, respectively. The sum over the column of the ρ_{ij} needs to satisfy $\sum_i \rho_{ij} = 0$ to ensure additivity. The parameter β_j locates the inflection point of the logistic at $\beta_j/2$. Finally, specification 22 corresponds to a Constant Elasticity of Transformation (CET) allocation across a row where $\sigma_i < 0$. If a firm is producing several outputs which are aggregated into a composite output represented by a CET, then the maximization of the revenues of the firm for a given level of the composite output will lead to a row allocation along specification 22. The parameters b_{ij} are such that $\sum_j b_{ij} = 1$ and and $0 < b_{ij} < 1$; they are the distribution parameters in the CET aggregation. In MISR2 specification 22 is used to allocate a given amount of workers remittances in foreign currency between two pools of foreign exchange depending on the relative exchange rates.

2. The Within-Period Module of the Reference Model of the MISR2 Class.

In part B of this chapter, the accounts of the MISR2 models were described. All the elements of the underlying SAM were identified. In the following we simply present the tables introduced in part B but with two modifications: (i) to each cell in the SAM one of the specifications of table 1 is assigned and (ii) the endogenous, exogenous or undefined nature of the price, value and quantity associated with each account is given. These two sets of information provide a complete information on the reference model.$\underline{3/}$

$\underline{1/}$ $\sigma_j > 0$ and $\sigma_j \neq 1$. If $\sigma_j = 1$ another specification (13) not used in MISR2 is needed.

$\underline{2/}$ Note that β_j does not depend on i and is the same across the entries in the column. Specification 16 is derived from a generalization proposed by Carlevaro (1976) of the Linear Expenditure System.

$\underline{3/}$ The TV software stores these two sets of information in two files: (i) TV-SAM file and (ii) an account-type file. For additional information on the interpretation of the tables, see Drud, Grais and Pyatt (1983).

Table 1: LIST OF SPECIFICATION USED IN THE MISR2 CLASS OF MODELS (*)

1. $t_{ij} = y_j$,

2. t_{ij} is a residual,

3. $t_{ij} = t_{ij}^o f_{ij}(\theta)$,

4. $t_{ij} = [\tau_j/(1 + \tau_j)]y_j$,

5. $t_{ij} = c_{ij} y_j$,

6. $t_{ij} = p_i t_{ij}^o f_{ij}(\theta)$,

7. $t_{ij} = t_{ij}^o f_{ij}(\theta) (p_j \pi_i^x)^{\eta_{ij}} (p_i)^{1-\eta_{ij}}$

8. $t_{ij} = t_{ij}^o f_{ij}(\theta)(p_j \pi_i^x)$,

9. t_{ij} is a residual and $p_i = \pi_i^x p_i$,

10. $t_{ij} = (t_{ij}^o/y_j^o) (p_i \pi_j^m/p_j) y_j$,

11. $t_{ij} = (t_{ij}^o/y_j^o) (p_i/p_j)^{1-\sigma_j} y_j$,

12. $t_{ij} = (t_{ij}^o/y_j^o)(p_i/p_j) y_j$,

15. $t_{ij} = a_{ij}(p_i/p_j) y_j$,

16. $t_{ij} = \{(t_{ij}^o/y_j^o) + \rho_{ij}[\exp(-\beta_j/(y_j/p_j)) - \exp(\beta_j/y_j^o)]\}y_j$,

19. $t_{ij} = \{(t_{ij}^o/y_j^o) + [f_{ij}(\theta)p_i/p_j]^{1-\sigma_j} y_j)\}$,

22. $t_{ij} = b_{ij}(p_i/p_j)^{\sigma-1} y_i$

(*) All variables with a superscript "o" are base year values drawn from the base year SAM.

TABLE A.2.1.
URBAN LABOR

					1	2	3	4	5	6	7	8	9	10	11	12	13	14	15	16	17	18	19	20	21
					LABOR ABROAD	URBAN	PUBLIC URBAN	PRIVATE URBAN	GOVERNMENT EMPLOYMENT	AGRICULTURE	FOOD PROCESSING	TEXTILES	OTHER INDUSTRIES	ELECTRICITY	CONSTRUCTION	OIL	TRANSPORTATION AND COMMUNICATIONS	SERVICES	AGRICULTURE	FOOD PROCESSING	TEXTILES	OTHER INDUSTRIES	CONSTRUCTION	TRANSPORTATION AND COMMUNICATIONS	SERVICES
										STATE ENTERPRISES									PRIVATE ENTERPRISES						
PRIMARY FACTORS — URBAN LABOR		URBAN	1	LABOR ABROAD																						
	CONSOLIDATED ACCOUNTS		2	URBAN																						
			3	PUBLIC URBAN		12			12																	
			4	PRIVATE URBAN		12				12	12	12	12	12	12	12	12	12	12	12	12	12	12	12	12
	CURRENT INSTITUTIONS ACCOUNTS		5	URBAN HSEHOLD REVENUE ACCOUNT	5	5	2	4	2	4	4	4	4	4	4	4	4	4	4	4	4	4	4	4	4
				REST OF WORLD 2 / COMMERCIAL BANKS		5																				
ACCOUNT TYPE				PRICE (P)	E	E	X	E	X	E	E	E	E	E	E	E	E	E	E	E	E	E	E	E	E
				VALUE (Y)	E	E	E	E	E	E	E	E	E	E	E	E	E	E	E	E	E	E	E	E	E
				QUANTITY (Q)	E	E	E	E	E	E	E	E	E	E	E	E	E	E	E	E	E	E	E	E	E

E -- endogenous
X -- exogenous
U -- undefined

TABLE A.2.2.
RURAL LABOR

				PRIMARY FACTORS — RURAL LABOR																		
					CONSOLIDATED ACCOUNTS				STATE ENTERPRISES								PRIVATE ENTERPRISES						
				LABOR ABROAD	RURAL	PUBLIC RURAL	PRIVATE RURAL	GOVERNMENT EMPLOYMENT	AGRICULTURE	FOOD PROCESSING	TEXTILES	OTHER INDUSTRIES	ELECTRICITY	CONSTRUCTION	OIL	TRANSPORTATION AND COMMUNICATIONS	SERVICES	AGRICULTURE	FOOD PROCESSING	TEXTILES	OTHER INDUSTRIES	CONSTRUCTION	TRANSPORTATION AND COMMUNICATIONS	SERVICES
				1	2	3	4	5	6	7	8	9	10	11	12	13	14	15	16	17	18	19	21	22
CUR-RENT INST. ACCT.	PRIMARY FACTORS RURAL LABOR	CONSOLIDATED ACCOUNT	RURAL	1																					
			PUBLIC RURAL	2	5	1	12		12																	
			PRIVATE RURAL	3		12																				
				4																					
		REVENUE ACCOUNT	RURAL HSEHOLD			2	2	4	2	4	4	4	4	4	4	4	4	4	4	4	4	4	4	4	4
ACCOUNT TYPE			PRICE (P)		E	X	X	E	X	E	E	E	E	E	E	EA	E	E	E	E	E	E	E	E	E
			VALUE (Y)		E	E	E	E	E	E	E	E	E	E	E	EA	E	E	E	E	E	E	E	E	E
			QUANTITY (Q)		X	E	E	E	E	E	E	E	E	E	E	EA	E	E	E	E	E	E	E	E	E

EA -- EMPTY ACCOUNT

TABLE A.2.3.
TOTAL LABOR

				STATE ENTERPRISES									PRIVATE ENTERPRISES						
			GOVERNMENT	AGRICULTURE	FOOD PROCESSING	TEXTILES	OTHER INDUSTRIES	ELECTRICITY	CONSTRUCTION	OIL	TRANSPORT AND COMMUNICATIONS	SERVICES	AGRICULTURE	FOOD PROCESSING	TEXTILES	OTHER INDUSTRIES	CONSTRUCTION	TRANSPORT AND COMMUNICATIONS	SERVICES
			1	2	3	4	5	6	7	8	9	10	11	12	13	14	15	16	17
PRIMARY FACTORS	URBAN LABOR	GOVERNMENT EMPLOYMENT	1	15																
		STATE ENTERPRISES AGRICULTURE	2		12															
		FOOD PROCESSING	3			12														
		TEXTILES	4				12													
		OTHER INDUSTRIES	5					12												
		ELECTRICITY	6						12											
		CONSTRUCTION	7							12										
		OIL	8								12									
		TRANSPORTATION AND COMMUNICATIONS	9									12								
		SERVICES	10										12							
		PRIVATE ENTERPRISES AGRICULTURE	11											15						
		FOOD PROCESSING	12												15					
		TEXTILES	13													15				
		OTHER INDUSTRIES	14														15			
		CONSTRUCTION	15															15		
		TRANSPORTATION AND COMMUNICATIONS	16																15	
		SERVICES	17																	15
	RURAL LABOR	GOVERNMENT EMPLOYMENT	18	15																
		STATE ENTERPRISES AGRICULTURE	19		12															
		FOOD PROCESSING	20			12														
		TEXTILES	21				12													
		OTHER INDUSTRIES	22					12												
		ELECTRICITY	23						12											
		CONSTRUCTION	24							12										
		OIL	25																	
		TRANSPORT AND COMMUNICATIONS	26									12								
		SERVICES	27										12							
		PRIVATE ENTERPRISES AGRICULTURE	28											15						
		FOOD PROCESSING	29												15					
		TEXTILES	30													15				
		OTHER INDUSTRIES	31														15			
		CONSTRUCTION	32															15		
		TRANSPORT AND COMMUNICATIONS	33																15	
		SERVICES	34																	15
	CAPITAL-PUBLIC	AGRICULTURE	35		2															
		FOOD PROCESSING	36			2														
		TEXTILES	37				2													
		OTHER INDUSTRIES	38					2												
		ELECTRICITY	39						2											
		CONSTRUCTION	40							2										
		TRANSPORT AND COMMUNICATIONS	41									2								
		SERVICES	42										2							
ACCOUNT TYPE		PRICE (P)		E	E	E	E	E	E	E	E	E	E	E	E	E	E	E	E	E
		VALUE (Y)		E	E	E	E	E	E	E	E	E	E	E	E	E	E	E	E	E
		QUANTITY (Q)		E	X	X	X	X	E	X	E	X	E	E	E	E	E	E	E	E

TABLE A.2.4.
CAPITAL AND LAND

			PRIMARY FACTORS																				
			CAPITAL																		LAND		
				PUBLIC									PRIVATE										
			NATIONAL CAPITAL	AGRICULTURE	FOOD PROCESSING	TEXTILES	OTHER INDUSTRIES	ELECTRICITY	CONSTRUCTION	OIL	TRANSPORT & COMMUNICATIONS	SERVICES	AGRICULTURE	FOOD PROCESSING	TEXTILES	OTHER INDUSTRIES	CONSTRUCTION	TRANSPORT & COMMUNICATIONS	SERVICES	CONSOLIDATED RENT	LAND 1 -- PUBLIC	LAND 2 -- PRIVATE
			1	2	3	4	5	6	7	8	9	10	11	12	13	14	15	16	17	18	19	20
CAPITAL		NATIONAL CAPITAL	1																				
LAND		CONSOLIDATED RENT	2																		1	1
HOUSE-HOLDS	URBAN	REVENUE ACCOUNT	3	5		5	5	5		5	5	5	5	5	5	5	5	5	5	5	5		
	RURAL	REVENUE ACCOUNT	4	5		5	5	5		5	5	5	5	5	5	5	5	5	5	5	5		
PRIVATE COMPANIES			5	5										5	5	5	5	5	5	5			
PUBLIC COM-PANIES		EGPC	6	5							5												
		OTHERS	7	5	1	5	5	5	1	5	5	5	5								5		
GOVERNMENT		GOVT. REVENUE	8	5								5											
ACCOUNT TYPE		PRICE (P)	U	E	E	E	E	E	E	E	E	E	E	E	E	E	E	E	E	U	E	E
		VALUE (Y)	E	E	E	E	E	E	E	E	E	E	E	E	E	E	E	E	E	E	E	E
		QUANTITY (Q)	U	X	X	X	X	X	X	X	X	X	X	X	X	X	X	X	X	U	X	X

TABLE A.2.5

COMPOSITE DOMESTIC INTERMEDIATE INPUTS: PUBLIC ACTIVITIES

					COMPOSITE DOMESTIC INTERMEDIATE INPUTS — PUBLIC ACTIVITIES										
					AGRICULTURE	TOTAL INTERMEDIATE	FOOD PROCESSING — COMPOSITE AGRICULTURE	FOOD PROCESSING — AGRICULTURE RENT ACCOUNT	TEXTILES	OTHER INDUSTRIES	ELECTRICITY	CONSTRUCTION	OIL	TRANSPORT AND COMMUNICATIONS	SERVICES
					1	2	3	4	5	6	7	8	9	10	11
DOM. GOVT. TR.CO		AGRICULTURE	SUBSIDIZED VALUE	1				12							
NON-GOVERNMENT COMMODITIES	COMPOSITE DOMESTIC	AGRICULTURE		2	12		11		12	12				12	12
		FOOD PROCESSING		3	12	12			12	12				12	12
		TEXTILES		4	12	12			12	12		12	12	12	12
		OTHER INDUSTRIES		5	12	12			12	12	12	12	12	12	12
		ELECTRICITY		6	12	12			12	12	12	12	12	12	12
		CONSTRUCTION		7	12	12			12	12	12	12	12	12	12
		OIL		8	12	12			12	12	12	12	12	12	12
		TRANSPORT AND COMMUNICATIONS		9	12	12			12	12	12	12	12	12	12
		SERVICES		10	12	12			12	12	12	12	12	12	12
COMPOSITE INPUTS	PUBLIC DOMESTIC	FOOD PROCESSING	AGRIC. RENT ACCOUNT	11			11								
			COMPOSITE AGRICULT.	12		12									
PRIM. FACT.	CAPITAL	PUBLIC	FOOD PROCESS.	13				2							
ACCOUNT TYPE		PRICE (P)			E	E	E	E	E	E	E	E	E	E	E
		VALUE (Y)			E	E	E	E	E	E	E	E	E	E	E
		QUANTITY (Q)			E	E	E	X	E	E	E	E	E	E	E

TABLE A.2.6

COMPOSITE DOMESTIC INTERMEDIATE INPUTS: PRIVATE ACTIVITIES

					COMPOSITE DOMESTIC INTERMEDIATE INPUTS — PRIVATE ACTIVITIES								
					AGRICULTURE	TOTAL INTERMEDIATE	FOOD PROCESSING — COMPOSITE AGRICULTURE	FOOD PROCESSING — AGRICULTURE RENT ACCOUNT	TEXTILES	OTHER INDUSTRIES	CONSTRUCTION	TRANSPORT AND COMMUNICATIONS	SERVICES
					1	2	3	4	5	6	7	8	9
DOM. GOVT. TRA. COM.		AGRICULTURE	SUBSIDIZED VALUE	1				12					
NON-GOVERNMENT COMMODITIES	COMPOSITE DOMESTIC	AGRICULTURE		2	I2		11		12	12		12	12
		FOOD PROCESSING		3	I2	12			12	12		12	12
		TEXTILES		4	I2	12			12	12	12	12	12
		OTHER INDUSTRIES		5	I2	12			12	12	12	12	12
		ELECTRICITY		6	12	12			12	12	12	12	12
		CONSTRUCTION		7	12	12			12	12	12	12	12
		OIL		8	12	12			12	12	12	12	12
		TRANSPORT AND COMMUNICATIONS		9	12	12			12	12	12	12	12
		SERVICES		10	12	12			12	12	12	12	12
COMPOSITE INPUTS	PRIVATE DOMESTIC FOOD PROCESSING		AGRICULTURE RENT ACCOUNT	11			11						
			COMPOSITE AGRICULTURE	12		12							
PRIMARY FACTORS	PRIVATE CAPITAL		FOOD PROCESSING	13				2					
ACCOUNT TYPE		PRICE (P)			E	E	E	E	E	E	E	E	E
		VALUE (Y)			E	E	E	E	E	E	E	E	E
		QUANTITY (Q)			E	E	E	X	E	E	E	E	E

TABLE A.2.7
COMPOSITE IMPORTED INTERMEDIATE INPUTS: PUBLIC ACTIVITIES

					1	2	3	4	5	6	7	8	9	10	11	12	13	14	15
					TOTAL INTERMEDIATE	COMPOSITE OTHER INDUSTRIES	OTHER INDUSTRIES RENT ACCOUNT	TOTAL INTERMEDIATE	COMPOSITE AGRICULTURE	COMPOSITE FOOD PROCESSING	AGRICULTURE RENT ACCOUNT	FOOD PROCESSING RENT ACCOUNT	TEXTILES	OTHER INDUSTRIES	ELECTRICITY	CONSTRUCTION	OIL	TRANSPORT AND COMMUNICATIONS	SERVICES
GOVERNMENT TRADE IMPORTS		AGRICULTURE	SUBSIDIZED VALUE	1							(12)								
		FOOD PROCESSING		2								(12)							
		OTHER INDUSTRIES		3			(12)												
NON-GOVERNMENT IMPORTS COMPOSITE IMPORTS		AGRICULTURE		4	(12)				(11)				(12)	(12)					(12)
		FOOD PROCESSING		5						(11)				(12)				(12)	(12)
		TEXTILES		6	(12)			(12)					(12)	(12)				(12)	(12)
		OTHER INDUSTRIES		7		(11)		(12)					(12)	(12)	(12)	(12)	(12)	(12)	(12)
		OIL		8	(12)			(12)					(12)	(12)	(12)	(12)	(12)	(12)	(12)
		TRANSPORT AND COMMUNICATIONS		9														(12)	
		SERVICES		10														(12)	(12)
COMPOSITE INPUTS	PUBLIC IMPORTS	AGRICULTURE	TOTAL INTERMEDIATE	11															
			OTHER INDUST. RENT ACCOUNT	12	(11)														
			COMPOSITE OTHER INDUST.	13	(12)														
PRIM. FACT.	CAPITAL		PUBLIC AGRICULTURE	14		(2)													
COMPOSITE INPUTS	PUBLIC IMPORTS FOOD PROCESSING		AGRICULTURE RENT ACCOUNT	15					(11)										
			FOOD PROCESS. RENT ACCOUNT	16						(11)									
			COMPOSITE AGRICULTURE	17				(12)											
			COMPOSITE FOOD PROCESS.	18				(12)											
PRIM. FACT.	PUBLIC CAPITAL		FOOD PROCESSING	19							(2)	(2)							
ACCOUNT TYPE		PRICE (P)			E	E	E	E	E	E	E	E	E	E	E	E	E	E	E
		VALUE (Y)			E	E	E	E	E	E	E	E	E	E	E	E	E	E	E
		QUANTITY (Q)			E	E	X	E	E	E	X	X	E	E	E	E	E	E	E

TABLE A.2.8
COMPOSITE IMPORTED INTERMEDIATE INPUTS: PRIVATE ACTIVITIES

				COMPOSITE INPUTS — IMPORTS / PRIVATE ACTIVITIES													
				AGRICULTURE					FOOD PROCESSING								
				TOTAL INTERMEDIATE	COMPOSITE AGRICULTURE	COMPOSITE OTHER INDUSTRIES	AGRICULTURE RENT ACCOUNT	OTHER INDUSTRIES RENT ACCOUNT	TOTAL INTERMEDIATE	COMPOSITE AGRICULTURE	COMPOSITE FOOD PROCESSING	AGRICULTURE RENT ACCOUNT	FOOD PROCESSING RENT ACCOUNT	TEXTILES	OTHER INDUSTRIES	CONSTRUCTION	TRANSPORT AND COMMUNICATIONS	SERVICES
				1	2	3	4	5	6	7	8	9	10	11	12	13	14	15
GOVERNMENT TRADE IMPORTS		AGRICULTURE	1				(12)					(12)						
		FOOD PROCESSING	2										(12)					
	SUBSIDIZED VALUE	OTHER INDUSTRIES	3					(12)										
NON-GOVERNMENT IMPORTS	COMPOSITE INPUTS	AGRICULTURE	4		(11)				(11)					(12)	(12)			(12)
		FOOD PROCESSING	5								(11)					(12)		(12)	(12)
		TEXTILES	6	(12)					(12)					(12)	(12)			(12)
		OTHER INDUSTRIES	7			(11)			(12)					(12)	(12)	(12)	(12)	(12)
		OIL	8	(12)					(12)					(12)	(12)	(12)	(12)	(12)
		TRANSPORT AND COMMUNICATIONS	9														(12)	
		SERVICES	10														(12)	(12)
COMPOSITE INPUTS	PRIVATE IMPORTS	AGRICULTURE RENT ACCOUNT	11		(11)													
		OTHER INDUSTRIES RENT ACCOUNT	12			(11)												
	AGRICULTURE	AGRICULTURE RENT TRANSFER	13															
		OTHER INDUSTRIES RENT TRANSFER	14															
		COMPOSITE AGRICULTURE	15	(12)														
		COMPOSITE OTHER INDUSTRIES	16	(12)														
PRIMARY FACTORS	CAPITAL	PRIVATE AGRICULTURE	17				(2)	(2)										
COMPOSITE INPUTS	PRIVATE IMPORTS	AGRICULTURE RENT ACCOUNT	18							(11)								
		FOOD PROCESSING RENT ACCOUNT	19								(11)							
	FOOD PROCESSING	AGRICULTURE RENT TRANSFER	20															
		FOOD PROCESSING RENT TRANSFER	21															
		COMPOSITE AGRICULTURE	22						(12)									
		COMPOSITE FOOD PROCESSING	23						(12)									
PRIMARY FACTORS	PRIVATE CAPITAL	FOOD PROCESSING	24									(2)	(2)					
ACCOUNT TYPE		PRICE (P)		E	E	E	E	E	E	E	E	E	E	E	E	E	E	E
		VALUE (Y)		E	E	E	E	E	ES	E	E	E	E	E	E	E	E	E
		QUANTITY (Q)		E	E	E	X	X	E	E	E	X	X	E	E	E	EE	E

TABLE A.2.9
HOUSEHOLD ACCOUNTS

				URBAN HOUSEHOLDS								RURAL HOUSEHOLDS							
				CURRENT ACCOUNT INSTITUTIONS				CONSTRAINED DEMAND				CURRENT ACCOUNT INSTITUTIONS				CONSTRAINED DEMAND			
								GOVERNMENT COMMODITIES								GOVERNMENT COMMODITIES			
				REVENUE ACCOUNT	DISPOSABLE INCOME	COMMITTED EXPENDITURES	DISCRETIONARY EXPENDITURES	AGRICULTURE	FOOD PROCESSING	OTHER INDUSTRIES	SERVICES	REVENUE ACCOUNT	DISPOSABLE INCOME	COMMITTED EXPENDITURE	DISCRETIONARY EXPENDITURES	AGRICULTURE	FOOD PROCESSING	OTHER INDUSTRIES	SERVICES
				1	2	3	4	5	6	7	8	9	10	11	12	13	14	15	16
HOUSEHOLDS	URBAN	REVENUE ACCOUNT	1									5							
	RURAL		2	5															
	PRIVATE COMPANIES		3	5								5							
	OTHER PUBLIC COMPANIES		4	5								5							
	SOCIAL SECURITY		5	5								5							
	GOVERNMENT REVENUE		6	5								5							
	DIRECT TAXES		7	5								5							
R.O.W.2	COMMERCIAL BANKS		8	5								5							
HOUSEHOLDS	URBAN	DISPOSABLE	9	5					2	2	2								
	RURAL	INCOME	10									5					2	2	2
	URBAN	COMMITTED	11		5														
	RURAL	EXPENDITURES	12										5						
	URBAN	DISCRETIONARY	13		2														
	RURAL	EXPENDITURES	14										2						
PRIVATE SAVINGS POOL			15		5								5						
URBAN	AGRICULTURE		16			6	16												
	FOOD PROCESSING		17			6	16												
	OTHER INDUSTRIES		18			6	16												
	SERVICES		19			6	16												
RURAL	AGRICULTURE		20											6	16				
	FOOD PROCESSING		21											6	16				
	OTHER INDUSTRIES		22											6	16				
	SERVICES		23											6	16				
GOVERNMENT COMPOSITE COMMODITIES HOUSEHOLD DEMAND	AGRICULTURE		24					12								12			
	FOOD PROCESSING		25						12								12		
	OTHER INDUSTRIES		26							12								12	
	SERVICES		27								12								12
NON-GOVERNMENT COMPOSITE COMMODITIES DOMESTIC	AGRICULTURE		28			6	16							6	16				
	FOOD PROCESSING		29			6	16							6	16				
	TEXTILES		30			6	16							6	16				
	OTHER INDUSTRIES		31			6	16							6	16				
	ELECTRICITY		32			6	16							6	16				
	OIL		33			6	16							6	16				
	TRANSPORT AND COMMUNICATIONS		34			6	16							6	16				
	SERVICES		35			6	16							6	16				
NON-GOVERNMENT COMPOSITE IMPORTS IMPORTED	AGRICULTURE		36			6	16							6	16				
	FOOD PROCESSING		37			6	16							6	16				
	TEXTILES		38			6	16							6	16				
	OTHER INDUSTRIES		39			6	16							6	16				
	OIL		40			6	16							6	16				
	TRANSPORT AND COMMUNICATIONS		41			6	16							6	16				
	SERVICES		42			6	16							6	16				
ACCOUNT TYPE	PRICE (P)			U	U	E	E	E	E	E	E	U	U	E	E	E	E	E	E
	VALUE (Y)			E	E	E	E	E	E	E	E	E	E	E	E	E	E	E	E
	QUANTITY (Q)			U	U	E	E	X	X	X	E	U	U	E	E	X	X	X	E

TABLE A.2.10
COMPANIES ACCOUNTS

				CURRENT ACCOUNT INSTITUTIONS		
				PRIVATE COMPANIES	PUBLIC COMPANIES	
					EGPC	OTHERS
				1	2	3
CURRENT ACCOUNT INSTITUTIONS	URBAN HOUSEHOLDS	REVENUE ACCOUNT	1	5	5	5
	RURAL HOUSEHOLDS	REVENUE ACCOUNT	2			5
	OTHER PUBLIC COMPANIES		3	5	5	
	PRIVATE COMPANIES		4			5
	PUBLIC COMPANIES EGPC		5			5
CAPITAL ACCOUNT	PRIVATE SAVINGS POOL		6	5		
	PUBLIC SAVINGS POOL		7		5	5
CURRENT ACCOUNT INSTITUTIONS	R.O.W. 1 CENTRAL BANK		8		5	5
	R.O.W. 2 COMMERCIAL BANK		9			5
	GOVERNMENT REVENUE		10		5	5
	SOCIAL SECURITY		11			5
	DIRECT TAXES		12	5	5	5
ACCOUNT TYPE	PRICE (P)			U	U	U
	VALUE (Y)			E	E	E
	QUANTITY (Q)			U	U	U

TABLE A.2.11
GOVERNMENT, SOCIAL SECURITY, AND THE TAX ACCOUNTS

						CURRENT ACCOUNT INSTITUTIONS						

					GOVERNMENT							SOCIAL SECURITY	TAXES				
						COMMITTED			DISCRETIONARY									
					GOVERNMENT REVENUE	EDUCATION	HEALTH	OTHER	EDUCATION	HEALTH	OTHER	GOVERNMENT TRADE	SOCIAL SECURITY	INDIRECT	SUBSIDY	DIRECT	IMPORT TARIFFS	EXPORT TAXES
					1	2	3	4	5	6	7	8	9	10	11	12	13	14
CURRENT ACCOUNT INSTITUTIONS	GOVERNMENT		REVENUE	1								1		1	1	1	1	1
		COMMITTED	EDUCATION	2	3													
			HEALTH	3	3													
			OTHER	4	3													
	HOUSEHOLDS	URBAN	REVENUE	5	3								3					
		RURAL	ACCOUNT	6	3								3					
	PUBLIC COMPANIES		EGPC	7	3													
			OTHERS	8	3								3					
	SOCIAL SECURITY			9	3													
CAPITAL ACCOUNT	PUBLIC SAVINGS POOL			10	2								2					
TOTAL LABOR	GOVERNMENT			11		6	6	6					3					
DOMESTIC GOVT. TRADE COMMODITIES	SUBSIDIZED VALUE	AGRICULTURE		12		6	6	6										
		FOOD PROCESSING		13		6	6	6										
GOVERNMENT TRADE IMPORTS		AGRICULTURE		14		6	6	6										
		FOOD PROCESSING		15		6	6	6										
CURRENT ACCOUNT INSTITUTIONS	GOVERNMENT DISCRETIONARY	EDUCATION		16	2													
		HEALTH		17					2									
		OTHER		18						2								
NON-GOVERNMENT COMMODITIES COMPOSITE DOMESTIC	AGRICULTURE			19					12	12	12							
	FOOD PROCESSING			20					12	12	12							
	TEXTILES			21					12	12	12							
	OTHER INDUSTRIES			22					12	12	12							
	ELECTRICITY			23					12	12	12							
	CONSTRUCTION			24					12	12	12							
	OIL			25					12	12	12							
	TRANSPORT AND COMMUNICATIONS			26					12	12	12							
	SERVICES			27					12	12	12							
NON-GOVERNMENT IMPORTS COMPOSITE IMPORTS	AGRICULTURE			28					12	12	12							
	FOOD PROCESSING			29					12	12	12							
	TEXTILES			30					12	12	12							
	OTHER INDUSTRIES			31					12	12	12							
	OIL			32					12	12	12							
	TRANSPORT AND COMMUNICATIONS			33					12	12	12							
	SERVICES			34					12	12	12							
CURRENT ACCOUNT INSTITUTIONS	REST OF THE WORLD 1	CENTRAL BANK		35	3													
ACCOUNT TYPE	PRICE (P)				U	E	E	E	E	E	E	U	U	U	U	U	U	U
	VALUE (Y)				E	E	E	E	E	E	E	E	E	E	E	E	E	E
	QUANTITY (Q)				U	E	E	E	E	E	E	U	U	U	U	U	U	U

Table A.2.12

Rest of the World Accounts

			CURRENT ACCOUNT INSTITUTIONS			
			REST OF WORLD 1	TRANSFER	REST OF WORLD 2	REST OF WORLD 3
			CENTRAL BANK	REST OF WORLD 1 AND 2	COMMERCIAL BANKS	OWN EXCHANGE
			1	2	3	4
PRIMARY FACTORS	LABOR ABROAD	1			㉒	㉒
	NATIONAL CAPITAL	2	⑧		⑧	
CURRENT ACCOUNT INSTITUTIONS	HOUSEHOLD REVENUE ACCOUNT — URBAN	3			⑧	
	HOUSEHOLD REVENUE ACCOUNT — RURAL	4			⑧	
	OTHER PUBLIC COMPANIES	5			⑧	
	GOVERNMENT REVENUE	6	⑧			
CAPITAL ACCOUNT	SAVINGS POOL — PUBLIC	7	⑧		⑧	
	SAVINGS POOL — PRIVATE	8				⑧
GOVT. TRADE EX.	PRIVATE ACTIV. — AGRICULTURE	9	⑦			
NON-GOVERNMENT COMMODITIES	COMPOSITE EXPORTS — AGRICULTURE	10	⑦			
	COMPOSITE EXPORTS — FOOD PROCESSING	11				⑦
	COMPOSITE EXPORTS — TEXTILES	12				⑦
	COMPOSITE EXPORTS — OTHER INDUSTRIES	13				⑦
	COMPOSITE EXPORTS — OIL	14	⑨			
	TRANSPORT AND COMM.S. — SUEZ CANAL	15	⑧			
	TRANSPORT AND COMM.S. — OTHER	16			⑦	
	SERVICES	17			⑦	
CURRENT ACCOUNT INSTITUTIONS	ROW 1 POOL — CENTRAL BANK	18		⑫		
	ROW 1 POOL — IMPORT PREMIA	19		②		
	TRANSFER — ROW 1 AND ROW 2	20			②	
ACCOUNT TYPE	PRICE (P)		X	X	X	E
	VALUE (Y)		E	E	E	E
	QUANTITY (Q)		E	E	E	E

TABLE A.2.13
CAPITAL ACCOUNTS

			CAPITAL ACCOUNT																					
			PRIVATE SAVINGS POOL	PRIVATE INVESTMENT	PUBLIC SAVINGS POOL	INVESTMENT		PUBLIC ACTIVITIES								PRIVATE ACTIVITIES							CAPITAL GOODS	
						GOVERNMENT	STATE ENTERPRISES	AGRICULTURE	FOOD PROCESSING	TEXTILES	OTHER INDUSTRIES	ELECTRICITY	CONSTRUCTION	OIL	TRANSPORT & COMM.	SERVICES	AGRICULTURE	FOOD PROCESSING	TEXTILES	OTHER INDUSTRIES	CONSTRUCTION	TRANSPORT & COMM.	SERVICES	PRIVATE SECTOR	PUBLIC SECTOR
			1	2	3	4	5	6	7	8	9	10	11	12	13	14	15	16	17	18	19	20	21	22	23
	PUBLIC SAVINGS POOL	1	5																							
	PRIVATE INVESTMENT	2	5																							
	GOVERNMENT INVESTMENT	3			2																					
	STATE ENTERPRISE INVEST	4			2																					
PUBLIC ACTIVITIES	AGRICULTURE	5					5																			
	FOOD PROCESSING	6					5																			
	TEXTILES	7					5																			
	OTHER INDUSTRIES	8					5																			
	ELECTRICITY	9					5																			
	CONSTRUCTION	10					5																			
	OIL	11					5																			
	TRANSPORT & COMM.	12					5																			
	SERVICES	13					5																			
PRIVATE ACTIVITIES	AGRICULTURE	14		5																						
	FOOD PROCESSING	15		5																						
	TEXTILES	16		5																						
	OTHER INDUSTRIES	17		5																						
	CONSTRUCTION	18		5																						
	TRANSPORT & COMM.	19		5																						
	SERVICES	20		5																						
CAPITAL GOODS	PRIVATE SECTOR	21															12	12	12	12	12	12	12		
	PUBLIC SECTOR	22						12	12	12	12	12	12	12	12	12									
DOM.GOVT.TR.CO. VALUE	AGRICULTURE	23				15																				
GOVT. TRADE IMP. SUBS.	AGRICULTURE	24				15																				
	FOOD PROCESSING	25				15																				
NON-GOVERNMENT COMMODITIES COMPOSITE DOMESTIC	AGRICULTURE	26				15																				15
	FOOD PROCESSING	27																								15
	TEXTILES	28																								15
	OTHER INDUSTRIES	29				15																			15	15
	ELECTRICITY	30																								
	CONSTRUCTION	31				15																			15	15
	OIL	32																								15
	TRANSPORT & COMM.	33				15																			15	15
	SERVICES	34				15																			15	15
NON-GOVERNMENT IMPORTS COMPOSITE IMPORTS	AGRICULTURE	35				15																			15	15
	FOOD PROCESSING	36																								15
	TEXTILES	37																								
	OTHER INDUSTRIES	38				15																			15	15
	OIL	39																								
	TRANSPORT & COMM.	40				15																			15	15
ACCOUNT TYPE	PRICE (P)		E	E	E	E	E	E	E	E	E	E	E	E	E	E	E	E	E	E	E	E	E	E	E	E
	VALUE (Y)		E	E	E	X	X	E	E	E	E	E	E	E	E	E	E	E	E	E	E	E	E	E	E	E
	QUANTITY (Q)		E	E	E	E	E	E	E	E	E	E	E	E	E	E	E	E	E	E	E	E	E	E	E	E

TABLE A.2.14
PUBLIC ACTIVITIES

			AGRICULTURE			FOOD PROCESSING			TEXTILES			OTHER INDUSTRIES			ELECTRICITY			CONSTRUCTION			OIL			TRANSPORT & COMMUNICATION			SERVICES		
			FACTOR COST	PRODUCERS COST	USER COST	FACTOR COST	PRODUCERS COST	USER COST	FACTOR COST	PRODUCERS COST	USER COST	FACTOR COST	PRODUCERS COST	USER COST	FACTOR COST	PRODUCERS COST	USER COST	FACTOR COST	PRODUCERS COST	USER COST	FACTOR COST	PRODUCERS COST	USER COST	FACTOR COST	PRODUCERS COST	USER COST	FACTOR COST	PRODUCERS COST	USER COST
			1	2	3	4	5	6	7	8	9	10	11	12	13	14	15	16	17	18	19	20	21	22	23	24	25	26
AGRICULTURE	LAND	1	19																									
	CAPITAL	2	19		4																							
	LABOR	3	19																									
	DOMESTIC COMP. INPUTS	4	19																									
	IMPORTED COMP. INPUTS	5	19																									
	FACTOR COST	6		12																								
	PRODUCERS COST	7			12																							
FOOD PROCESSING	CAPITAL	8				19		4																				
	LABOR	9				19																						
	DOMESTIC COMP. INPUTS	10				19																						
	IMPORTED COMP. INPUTS	11				19																						
	FACTOR COST	12					12																					
	PRODUCERS COST	13						12																				
TEXTILES	CAPITAL	14							19		4																	
	LABOR	15							19																			
	DOMESTIC COMP. INPUTS	16							19																			
	IMPORTED COMP. INPUTS	17							19																			
	FACTOR COST	18								12																		
	PRODUCERS COST	19									12																	
OTHER INDUSTRIES	CAPITAL	20										19		4														
	LABOR	21										19																
	DOMESTIC COMP. INPUTS	22										19																
	IMPORTED COMP. INPUTS	23										19																
	FACTOR COST	24											12															
	PRODUCERS COST	25												12														
ELECTRICITY	CAPITAL	26													19		2											
	LABOR	27													19													
	DOMESTIC COMP. INPUTS	28													19													
	IMPORTED COMP. INPUTS	29													19													
	FACTOR COST	30														12												
	PRODUCERS COST	31															12											
CONSTRUCTION	CAPITAL	32																19		4								
	LABOR	33																19										
	DOMESTIC COMP. INPUTS	34																19										
	IMPORTED COMP. INPUTS	35																19										
	FACTOR COST	36																	12									
	PRODUCERS COST	37																		12								
OIL	CAPITAL	38																			2							
	LABOR	39																			12							
	DOMESTIC COMP. INPUTS	40																			12							
	IMPORTED COMP. INPUTS	41																			12							
	FACTOR COST	42																				12						
	PRODUCERS COST	43																					19		4			
TRANSPORT AND COMMUNICATIONS	CAPITAL	44																					19					
	LABOR	45																					19					
	DOMESTIC COMP. INPUTS	46																					19					
	IMPORTED COMP. INPUTS	47																						12				
	FACTOR COST	48																							12			
	PRODUCERS COST	49																								19		2
SERVICES	CAPITAL	50																								19		
	LABOR	51																								19		
	DOMESTIC COMP. INPUTS	52																								19		
	IMPORTED COMP. INPUTS	53																									12	
	FACTOR COST	54																										12
	PRODUCERS COST	55																										
CURRENT ACCOUNT INSTITUTIONS INDIRECT TAXES		56	4			4			4			4			4			4			4			4			4		
ACCOUNT TYPE	PRICE (P)		X	X	X	X	X	X	X	X	X	X	X	X	X	X	X	X	X	X	X	X	X	X	X	X	X	X
	VALUE (Y)		X	X	X	X	X	X	X	X	X	X	X	X	X	X	X	X	X	X	X	X	X	X	X	X	X	X
	QUANTITY (Q)		X	X	X	X	X	X	X	X	X	X	X	X	X	X	X	X	X	X	X	X	X	X	X	X	X	X

TABLE A.2.15

PRIVATE ACTIVITIES

			PRIVATE ACTIVITIES												
			AGRI-CULTURE		FOOD PROCESS.		TEX-TILES		OTHER INDUST.		CONS-TRUCT.		TRANSP. & COMM.		SER-VICES	
			FACTOR COST	PRODUCERS COST	FACTOR COST	PRODUCERS COST	FACTOR COST	PRODUCERS COST	FACTOR COST	PRODUCERS COST	FACTOR COST	PRODUCERS COST	FACTOR COST	PRODUCERS COST	FACTOR COST	PRODUCERS COST
			1	2	3	4	5	6	7	8	9	10	11	12	13	14
AGRICULTURE	LAND	1	19													
	CAPITAL	2	19													
	LABOR	3	19													
	DOMESTIC COMP. INPUTS	4	19													
	IMPORTED COMP. INPUTS	5	19													
	FACTOR COST	6		12												
FOOD PROCESSING	CAPITAL	7			19											
	LABOR	8			19											
	DOMESTIC COMP. INPUTS	9			19											
	IMPORTED COMP. INPUTS	10			19											
	FACTOR COST	11				12										
TEXTILES	CAPITAL	12					19									
	LABOR	13					19									
	DOMESTIC COMP. INPUTS	14					19									
	IMPORTED COMP. INPUTS	15					19									
	FACTOR COST	16						12								
OTHER INDUSTRIES	CAPITAL	17							19							
	LABOR	18							19							
	DOMESTIC COMP. INPUTS	19								19						
	IMPORTED COMP. INPUTS	20								19						
	FACTOR COST	21									12					
CONSTRUCTION	CAPITAL	22									19					
	LABOR	23									19					
	DOMESTIC COMP. INPUTS	24									19					
	IMPORTED COMP. INPUTS	25									19					
	FACTOR COST	26											12			
TRANSPORT & COMMUNICATIONS	CAPITAL	27											19			
	LABOR	28											19			
	DOMESTIC COMP. INPUTS	29											19			
	IMPORTED COMP. INPUTS	30											19			
	FACTOR COST	31												12		
SERVICES	CAPITAL	32													19	
	LABOR	33													19	
	DOMESTIC COMP. INPUTS	34													19	
	IMPORTED COMP. INPUTS	35													19	
	FACTOR COST	36														12
CURRENT ACCOUNT INSTITUTIONS	INDIRECT TAXES	37		4		4		4		4		4		4		4
ACCOUNT TYPE	PRICE (P)		E	E	E	E	E	E	E	E	E	E	E	E	E	E
	VALUE (Y)		E	E	E	E	E	E	E	E	E	E	E	E	E	E
	QUANTITY (Q)		E	E	E	E	E	E	E	E	E	E	E	E	E	E

TABLE A.2.16

GOVERNMENT TRADE COMMODITIES (DOMESTIC, IMPORTS AND EXPORTS)

				GOVERNMENT COMPOSITE COMMODITIES				DOMESTIC GOVERNMENT TRADE COMMODITIES							GOVERNMENT TRADE IMPORTS						GOVT. TRADE EXPORTS		
				HOUSEHOLD DEMAND				AGRICULTURE		FOOD PROCESSING					AGRICULTURE		FOOD PROCESS.		OTHER INDUST.		AGRICULTURE		
				AGRICULTURE	FOOD PROCESSING	OTHER INDUSTRIES	SERVICES	FIXED PRICE TO PRODUCER	COST TO GOVERNMENT	SUBSIDIZED VALUE	FULL VALUE - PUBLIC	FULL VALUE - PRIVATE	SUBSIDIZED COMPOSITE	SERVICES	LANDED VALUE	SUBSIDIZED VALUE	LANDED VALUE	SUBSIDIZED VALUE	LANDED VALUE	SUBSIDIZED VALUE	FIXED PRICE TO PRODUCER	COST TO GOVERNMENT	SUPPLY PRICE
				1	2	3	4	5	6	7	8	9	10	11	12	13	14	15	16	17	18	19	20
DOMESTIC GOVERNMENT TRADE COMMODITIES	SUBSIDIZED VALUE	AGRICULTURE	1	6																			
		FOOD PROCESSING	2		6																			
		SERVICES	3				12																
GOVERNMENT TRADE IMPORTS	SUBSIDIZED VALUE	AGRICULTURE	4	2																			
		FOOD PROCESSING	5		2																			
		OTHER INDUSTRIES	6			12																		
ACTIVITY	PUBLIC	USER COST	FOOD PROCESSING	7								12												
			SERVICES	8								4												
	PRIVATE	PRODUCERS COST	AGRICULTURE	9					12													12		
			FOOD PROCESSING	10										12											
			SERVICES	11							4			4									2		
DOMESTIC GOVERNMENT TRADE COMMODITIES	AGRICULTURE	FIXED PRICE TO PRODUCER	12					12																
		COST TO GOVERNMENT	13						12															
	FULL VALUE	FOOD PROCESSING - PUBLIC	14								12													
		FOOD PROCESSING - PRIVATE	15									12												
CURRENT ACCOUNT INSTITUTIONS		GOVERNMENT TRADE	16						2		2	1			2		2		2				
	R.O.W.1	CENTRAL BANK	17													10		10		10				
		IMPORT TARIFFS	18													4		4		4				
GOVERNMENT TRADE IMPORTS	LANDED VALUE	AGRICULTURE	19												12									
		FOOD PROCESSING	20														12							
		OTHER INDUSTRIES	21																12					
GOVERNMENT TRADE EXPORTS	AGRICULTURE	FIXED PRICE TO PRODUCER	22																			12		
		COST TO GOVERNMENT	23																				12	
PRIMARY FACTORS	CAPITAL	PRIVATE AGRICULTURE	24					2														2		
CURRENT ACC. INSTITUTIONS		EXPORT TAXES	25																					2
ACCOUNT TYPE		PRICE (P)		E	E	E	E	X	E	X	E	E	E	X	X	E	X	E	X	E	X	E	E
		VALUE (Y)		E	E	E	E	E	E	E	E	E	E	E	E	E	E	E	E	E	E	E	E
		QUANTITY (Q)		E	E	E	E	E	E	E	E	E	E	E	E	E	E	E	E	E	E	E	X

TABLE A.2.17
NON-GOVERNMENT COMMODITIES (DOMESTIC AND EXPORTS)

TABLE A.2.18

NON-GOVERNMENT COMPOSITE COMMODITIES (DOMESTIC AND EXPORTS)

			NON-GOVERNMENT COMMODITIES															
			COMPOSITE DOMESTIC									COMPOSITE EXPORTS						
			AGRICULTURE	FOOD PROCESSING	TEXTILES	OTHER INDUSTRIES	ELECTRICITY	CONSTRUCTION	OIL	TRANSPORT & COMMUNICATIONS	SERVICES	AGRICULTURE	FOOD PROCESSING	TEXTILES	OTHER INDUSTRIES	OIL	TRANSPORT AND COMMUNICATIONS	SERVICES
			1	2	3	4	5	6	7	8	9	10	11	12	13	14	15	16
CURRENT ACCOUNT INSTITUTIONS	INDIRECT TAXES	1	4	4	4	4	4	4	4	4	4							
	SUBSIDIES	2		4	4		4	4	4	4								
PRIMARY FACTORS	PUBLIC CAPITAL	3														2		
NON-GOVERNMENT / DOMESTIC PUBLIC RENT	AGRICULTURE	4	11															
	FOOD PROCESSING	5		11														
	TEXTILES	6			11													
	OTHER INDUSTRIES	7				11												
	ELECTRICITY	8					12											
	CONSTRUCTION	9						11										
	OIL	10							12									
	TRANSPORT AND COMMUNICATIONS	11								11								
	SERVICES	12									11							
DOMESTIC PRIVATE	AGRICULTURE	13	11															
	FOOD PROCESSING	14		11														
	TEXTILES	15			11													
	OTHER INDUSTRIES	16				11												
	CONSTRUCTION	17						11										
	TRANSPORT AND COMMUNICATIONS	18								11								
	SERVICES	19									11							
PUBLIC EXPORTS	AGRICULTURE	20										11						
	FOOD PROCESSING	21											11					
	TEXTILES	22												11				
	OTHER INDUSTRIES	23													11			
	OIL	24														12		
	TRANSPORT AND COMMUNICATIONS	25															11	
	SERVICES	26																11
PRIVATE EXPORTS	AGRICULTURE	27										11						
	FOOD PROCESSING	28											11					
	TEXTILES	29												11				
	OTHER INDUSTRIES	30													11			
	TRANSPORT AND COMMUNICATIONS	31															11	
	SERVICES	32																11
ACCOUNT TYPE	PRICE (P)		E	E	E	E	E	E	E	E	E	E	E	E	E	E	E	E
	VALUE (Y)		E	E	E	E	E	E	E	E	E	E	E	E	E	E	E	E
	QUANTITY (Q)		E	E	E	E	E	E	E	E	E	E	E	E	E	E	E	E

TABLE A.2.19
NON-GOVERNMENT IMPORTS

Chapter 3 - The Alternative MISR2 Model: Changes in Policy Regime

1. Introduction

The MISR2 reference model captures the institutional arrangements and policy regime prevalent in Egypt around the middle of 1983. In particular it postulates fixed output prices for all public sector activities. It also assumes that trade and other transactions with the rest of the world pass through one of three foreign exchange pools: (i) the central bank; (ii) the commercial banks and (iii) the "parallel" market. The first two pools have a fixed exchange rate while the third pool has a flexible one.

One exercise undertaken with the MISR2 is to analyse the medium run macroeconomic consequences of a reform of the public sector pricing, trade, and exchange rate and employment regimes. The reform is designed to be implemented over a period of between five and ten years. During the first five years, the policy regime of 1983 is maintained but controlled adjustments in prices, and macroeconomic variables are implemented. The purpose is to gradually increase public sector prices in order to reduce distortions. Prices remain however policy instruments. Beyond the fifth year, under the presumption that the economy has acquired flexibility and agents are now more familiar with price changes, public sector prices and employment decisions are freed; proves either respond to excess demands or, like oil prices, follow the world price; public sector enterprises are free to hire and fire labor at the form of wage rate. Simultaneously to the freeing of public sector prices and employment decisions, transactions with the rest of the world are consolidated into two pools, instead of three, and the exchange rate on the second pool is left free to adjust. Thus the commercial banks and "parallel markets of foreign exchange are consolidated into one market with a free exchange rate. The central bank is allowed to buy and sell foreign exchange from the free market. One distortion remains there because of the difference in exchange rates governing transactions on the two pools.

This chapter briefly outlines how, the alternative MISR2 model, corresponding to the reformed economy differs from the reference model. We consider first the changes in specifications pertaining to public sector prices and then those corresponding to the reform of the trade and exchange rate regime.

2. Liberalizing Public Sector Output Prices

Tables A.3.14 and A.3.17 and A.3.18 provide the TV specificatons when public sector prices are free to adjust to excess demand and oil prices are determined by world prices. First, at the bottom of table A.3.14, the account types are changed and all prices are now specified to be endogenous. Corresponding to the endogenization of the nine prices, one needs to "exogenize" nine other variables in order to keep the model square. The natural variables to exogenize and to put identical to zero are the rents due to the distortions. This is done for the demand-driven sectors, electricity and services by assigning to the cells (26, 15) and (49, 26) the specification 4 instead of 2 (see table A.3.14). Specification 4 indicates the existence of exogenous mark-up between the market price and the critical price. By driving the mark-up to zero, the rents vanish. Along the same idea the specifications 2 in row 26 of table (A.3.17) are changed to 4. This latter change pertains

TABLE A.3.14
PUBLIC ACTIVITIES

TABLE A.3.17
NON-GOVERNMENT COMMODITIES (DOMESTIC AND EXPORTS)

to the supply-driven sectors. Finally for the oil sector cell (3,14) in table (A.3.18) is also changed to 4; hence the domestic and world prices of oil will stand in fixed proportions and the wedge can be made to vanish exogenously.

The foregoing changes, with their implications in terms of exogenous parameters to be given to the model, change the MISR2 model into one where all public sector prices are market determined.

3. Public Sector Employment

Another reform of the public sector is the liberalization of the hiring of labor decisions. The envisaged reform would allow state enterprises to hire and fire freely labor at the going wage rate. The implied change in specifications is shown in table A.3.3. On the one hand the quantities associated with column 2 to 10 become cudogenous reflecting the fact that employment decisions result from the behavior of enterprises and are no longer fixed exogenously. Simultaneously, the negative residual rents are transformed into exogenous rent by using specification 4 in the bottom of the table. Making these rent vanish would completely remove the distortions introduced by the previous fixity of labor.

4. Reforming the Exchange Rate and Trade Regime

The envisaged reform involves consolidating the commercial banks and "parallel" foreign exchange markets into one market with a flexible exchange rate. In so doing, the rationing of imports on the commercial banks pool is removed. The central bank pool which still has a fixed exchange rate and fixed net borrowing from abroad clears via purchasing and selling foreign exchange from the new consolidated market, called free market in the following.

In terms of foreign exchange receipts and transactions between the central bank and the free market, the flows and specifications are as shown in table (A.3.12). Now all workers remittances come through the free market as well as the net foreign borrowing of the private sector (direct investment). Similarly the non-agricultual commodity exports pass through the same market as the exports of services and transportation and communication. The central bank buys foreign exchange from the transfer account at its own exchange rate. The transfer account buys the foreign exchange from the free market at the free exchange rate and channels the difference to the import premia account. Corresponding to these changes the free market has an endogenous price--flexible exchange rate--while the central bank and the transfer account have the same exogenous price which is the fixed central bank exchange rate. This is shown in the account types appearing at the bottom of the table.

The changes in terms of imports are shown in table (A.3.19). The composite imports (columns 31 to 37) which used to come through three channels, come now through two; the central bank imports and the imports premia account. The former are treated as previously. The latter (in columns 17 to 23) buy now imports directly from the rest of the world (specifications 10 in row 7). Again here, like with the rents for public sector prices, the premia are exogenized using specification 4 in row 5. As the own-exchange market and the discretionary foreign exchange budget disappear; the accounts

TABLE A.3.18
NON-GOVERNMENT COMPOSITE COMMODITIES (DOMESTIC AND EXPORTS)

			COMPOSITE DOMESTIC									COMPOSITE EXPORTS							
			AGRICULTURE	FOOD PROCESSING	TEXTILES	OTHER INDUSTRIES	ELECTRICITY	CONSTRUCTION	OIL	TRANSPORT & COMMUNICATIONS	SERVICES	AGRICULTURE	FOOD PROCESSING	TEXTILES	OTHER INDUSTRIES	OIL	TRANSPORT AND COMMUNICATIONS	SERVICES
			1	2	3	4	5	6	7	8	9	10	11	12	13	14	15	16
CURRENT ACCOUNT INSTITUTIONS	INDIRECT TAXES	1	4	4	4	4	4	4	4	4	4							
	SUBSIDIES	2			4	4		4	4	4	4							
PRIMARY FACTORS	PUBLIC CAPITAL	3														4		
NON-GOVERNMENT / DOMESTIC PUBLIC RENT	AGRICULTURE	4	11															
	FOOD PROCESSING	5		11														
	TEXTILES	6			11													
	OTHER INDUSTRIES	7				11												
	ELECTRICITY	8					12											
	CONSTRUCTION	9						11										
	OIL	10							12									
	TRANSPORT AND COMMUNICATIONS	11								11								
	SERVICES	12									11							
DOMESTIC PRIVATE	AGRICULTURE	13	11															
	FOOD PROCESSING	14		11														
	TEXTILES	15			11													
	OTHER INDUSTRIES	16				11												
	CONSTRUCTION	17						11										
	TRANSPORT AND COMMUNICATIONS	18								11								
	SERVICES	19									11							
PUBLIC EXPORTS	AGRICULTURE	20										11						
	FOOD PROCESSING	21											11					
	TEXTILES	22												11				
	OTHER INDUSTRIES	23													11			
	OIL	24														12		
	TRANSPORT AND COMMUNICATIONS	25															11	
	SERVICES	26																11
PRIVATE EXPORTS	AGRICULTURE	27										11						
	FOOD PROCESSING	28											11					
	TEXTILES	29												11				
	OTHER INDUSTRIES	30													11			
	TRANSPORT AND COMMUNICATIONS	31															11	
	SERVICES	32																11
ACCOUNT TYPE	PRICE (P)		E	E	E	E	E	E	E	E	E	E	E	E	E	E	E	E
	VALUE (Y)		E	E	E	E	E	E	E	E	E	E	E	E	E	E	E	E
	QUANTITY (Q)		E	E	E	E	E	E	E	E	E	E	E	E	E	E	E	E

- 185 -

TABLE A.3.3.
TOTAL LABOR

				TOTAL LABOR																
				STATE ENTERPRISES										PRIVATE ENTERPRISES						
				GOVERNMENT	AGRICULTURE	FOOD PROCESSING	TEXTILES	OTHER INDUSTRIES	ELECTRICITY	CONSTRUCTION	OIL	TRANSPORT AND COMMUNICATIONS	SERVICES	AGRICULTURE	FOOD PROCESSING	TEXTILES	OTHER INDUSTRIES	CONSTRUCTION	TRANSPORT AND COMMUNICATIONS	SERVICES
				1	2	3	4	5	6	7	8	9	10	11	12	13	14	15	16	17
PRIMARY FACTORS	URBAN LABOR		GOVERNMENT EMPLOYMENT	1	15																
		STATE ENTERPRISES	AGRICULTURE	2		12															
			FOOD PROCESSING	3			12														
			TEXTILES	4				12													
			OTHER INDUSTRIES	5					12												
			ELECTRICITY	6						12											
			CONSTRUCTION	7							12										
			OIL	8								12									
			TRANSPORTATION AND COMMUNICATIONS	9									12								
			SERVICES	10										12							
		PRIVATE ENTERPRISES	AGRICULTURE	11											15						
			FOOD PROCESSING	12												15					
			TEXTILES	13													15				
			OTHER INDUSTRIES	14														15			
			CONSTRUCTION	15															15		
			TRANSPORTATION AND COMMUNICATIONS	16																15	
			SERVICES	17																	15
	RURAL LABOR		GOVERNMENT EMPLOYMENT	18	15																
		STATE ENTERPRISES	AGRICULTURE	19		12															
			FOOD PROCESSING	20			12														
			TEXTILES	21				12													
			OTHER INDUSTRIES	22					12												
			ELECTRICITY	23						12											
			CONSTRUCTION	24							12										
			OIL	25																	
			TRANSPORT AND COMMUNICATIONS	26									12								
			SERVICES	27										12							
		PRIVATE ENTERPRISES	AGRICULTURE	28											15						
			FOOD PROCESSING	29												15					
			TEXTILES	30													15				
			OTHER INDUSTRIES	31														15			
			CONSTRUCTION	32															15		
			TRANSPORT AND COMMUNICATIONS	33																15	
			SERVICES	34																	15
	CAPITAL-PUBLIC		AGRICULTURE	35		4															
			FOOD PROCESSING	36			4														
			TEXTILES	37				4													
			OTHER INDUSTRIES	38					4												
			ELECTRICITY	39						4											
			CONSTRUCTION	40							4										
			TRANSPORT AND COMMUNICATIONS	41									4								
			SERVICES	42										4							
ACCOUNT TYPE			PRICE (P)		E	E	E	E	E	E	E	E	E	E	E	E	E	E	E	E	E
			VALUE (Y)		E	E	E	E	E	E	E	E	E	E	E	E	E	E	E	E	E
			QUANTITY (Q)		E	E	E	E	E	E	E	E	E	E	E	E	E	E	E	E	E

Table A.3.12

Rest of the World Accounts

				CURRENT ACCOUNT INSTITUTIONS			
				REST OF WORLD 1	TRANSFER	REST OF WORLD 2	REST OF WORLD 3
				CENTRAL BANK	REST OF WORLD 1 AND 2	COMMERCIAL BANKS	OWN EXCHANGE
				1	2	3	4
PRIMARY FACTORS	LABOR ABROAD		1			(8)	
	NATIONAL CAPITAL		2	(8)		(8)	
CURRENT ACCOUNT INSTITUTIONS	HOUSEHOLD REVENUE ACCOUNT	URBAN	3			(8)	
		RURAL	4			(8)	
	OTHER PUBLIC COMPANIES		5			(8)	
	GOVERNMENT REVENUE		6	(8)			
CAPITAL ACCOUNT	SAVINGS POOL	PUBLIC	7	(8)		(8)	
		PRIVATE	8			(8)	
GOVT. TRADE EX.	PRIVATE ACTIV.	AGRICULTURE	9	(7)			
NON-GOVERNMENT COMMODITIES	COMPOSITE EXPORTS	AGRICULTURE	10	(7)			
		FOOD PROCESSING	11			(7)	
		TEXTILES	12			(7)	
		OTHER INDUSTRIES	13			(7)	
		OIL	14	(9)			
		TRANSPORT AND COMM.S. / SUEZ CANAL	15	(8)			
		TRANSPORT AND COMM.S. / OTHER	16			(7)	
		SERVICES	17			(7)	
CURRENT ACCOUNT INSTITUTIONS	ROW 1 POOL	CENTRAL BANK	18				
		IMPORT PREMIA	19		(2)		
	TRANSFER	ROW 1 AND ROW 2	20	(2)			
	ROW 2	COMMERCIAL BANKS + OWN EXCHANGE				(12)	
ACCOUNT TYPE	PRICE (P)			X		X	
	VALUE (Y)			E	E	E	
	QUANTITY (Q)			E	E	E	

TABLE A.3.19
NON-GOVERNMENT IMPORTS

				CURRENT ACCOUNT INST. REST OF WORLD		NON-GOVERNMENT IMPORTS																															
						CENTRAL BANK							DISCRETIONARY FOREIGN EXCHANGE BUDGET							IMPORTS PREMIA							OWN EXCHANGE							COMPOSITE IMPORTS					
				DISCRETIONARY FOREIGN EXCHANGE BUDGET	IMPORT PREMIA	AGRICULTURE	FOOD PROCESSING	TEXTILES	OTHER INDUSTRIES	OIL	TRANSPORT AND COMMUNICATIONS	SERVICES	AGRICULTURE	FOOD PROCESSING	TEXTILES	OTHER INDUSTRIES	OIL	TRANSPORT AND COMMUNICATIONS	SERVICES	AGRICULTURE	FOOD PROCESSING	TEXTILES	OTHER INDUSTRIES	OIL	TRANSPORT AND COMMUNICATIONS	SERVICES	AGRICULTURE	FOOD PROCESSING	TEXTILES	OTHER INDUSTRIES	OIL	TRANSPORT AND COMMUNICATIONS	SERVICES	AGRICULTURE	FOOD PROCESSING	TEXTILES	OTHER INDUSTRIES	OIL	TRANSPORT AND COMMUNICATIONS	SERVICES
				1	2	3	4	5	6	7	8	9	10	11	12	13	14	15	16	17	18	19	20	21	22	23	24	25	26	27	28	29	30	31	32	33	34	35	36	37
CURRENT ACCOUNT INSTITUTIONS	HOUSEHOLDS	REVENUE ACCOUNT	URBAN	1		5																																		
			RURAL	2		5																																		
	OTHER PUBLIC COMPANIES			3		5																																		
	R.O.W. DISCRETIONARY	FOREIGN EXCHANGE BUDGET		4																																				
	R.O.W. POOL	IMPORT PREMIA		5						2		2	2									4	4	4	4	4	4	4													
	R.O.W.1	CENTRAL BANK		6						10	10	10	10																												
	R.O.W.2	COMMERCIAL BANKS		7																	10	10	10	10	10	10	10													
	R.O.W.3	OWN EXCHANGE		8																																				
	TAXES	IMPORT TARIFFS		9																																4	4	4	4	4	4	4
NON-GOVERNMENT IMPORTS	CENTRAL BANK	AGRICULTURE		10																																					
		FOOD PROCESSING		11																																					
		TEXTILES		12																																					
		OTHER INDUSTRIES		13																																		11			
		ELECTRICITY		14																																					
		CONSTRUCTION		15																																					
		OIL		16																																			11		
		TRANSPORT AND COMMUNICATIONS		17																																				11	
		SERVICES		18																																					11
	DISCRETIONARY FOREIGN EXCHANGE BUDGET	AGRICULTURE		19																																					
		FOOD PROCESSING		20																																					
		TEXTILES		21																																					
		OTHER INDUSTRIES		22																																					
		ELECTRICITY		23																																					
		CONSTRUCTION		24																																					
		OIL		25																																					
		TRANSPORT AND COMMUNICATIONS		26																																					
		SERVICES		27																																					
	IMPORTS PREMIA	AGRICULTURE		28																																11					
		FOOD PROCESSING		29																																		11				
		TEXTILES		30																																			11			
		OTHER INDUSTRIES		31																																				11		
		ELECTRICITY		32																																					
		CONSTRUCTION		33																																					
		OIL		34																																				11		
		TRANSPORT AND COMMUNICATIONS		35																																					11	
		SERVICES		36																																						11
	OWN EXCHANGE	AGRICULTURE		37																																					
		FOOD PROCESSING		38																																					
		TEXTILES		39																																					
		OTHER INDUSTRIES		40																																					
		ELECTRICITY		41																																					
		CONSTRUCTION		42																																					
		OIL		43																																					
		TRANSPORT AND COMMUNICATIONS		44																																					
		SERVICES		45																																					
ACCOUNT TYPE		PRICE (P)		EA	U	EA	EA	EA	E	E	E	X	EA	EA	EA	EA	EA	EA	EA	E	E	E	E	E	E	E	EA	EA	EA	EA	EA	EA	EA	E	E	E	E	E	E	E
		VALUE (Y)		EA	E	EA	EA	EA	E	E	E	E	EA	EA	EA	EA	EA	EA	EA	E	E	E	E	E	E	E	EA	EA	EA	EA	EA	EA	EA	E	E	E	E	E	E	E
		QUANTITY (Q)		EA	U	EA	EA	EA	X	R	X	X	EA	EA	EA	EA	EA	EA	EA	E	E	E	E	E	E	E	EA	EA	EA	EA	EA	EA	EA	E	E	E	E	E	E	E

pertaining to them are empty. The corresponding exogenous or endogenous nature of prices and quantities is shown at the bottom of the table.

With the foregoing changes in the behavior of public sector prices and in the regime governing the transactions with the rest of the world, another model in the MISR2 class is obtained. It corresponds to an economy where there are less regulations and where distortions are less widespread.

APPENDIX 2

APPENDIX 2

The Inter-Period Module of the MISR2 Class of Models

Page No.

INTRODUCTION

CHAPTER 1: The Economics of Inter-Period Module of the Reference Model 192

   1. Resource Endowments 192

   2. Policy Measures 193

   3. The Rest of the World 194

CHAPTER 2: The Formulation of the Inter-Period Module of MISR2 194

   1. Fixed Capital Formation and Capital Stock Accumulation ... 194

   2. Rural-Urban Migration and Urban Wage Determination 197

CHAPTER 1: The Inter-Period Module of the Reference Model

MISR2 is a framework for policy analysis which allows to analyze the change in the general equilibrium of the economy over a sequence of periods, each of them covering a year. Thus, the framework can provide alternative paths of the economy as a sequence of equilibria. The previous appendix discussed how equilibrium in each period is defined. The linkages between the equilibria in the various periods are taken up here.

In each period equilibrium is obtained under specific assumptions on resource endowments (land, labor, capital), policy measures and trends in the rest of the world. It is the modification of these assumptions which shift the equilibrium from period to period.

1. **Resource Endowments**

Consider first the change in resource endowments. Land is a factor of production intervening in both public and private agriculture. Expansion of land means a potential for the expansion of the supply of agricultural goods. As is well known, land is a very scarce factor in Egypt. Its availability changes because of two phenomena: (i) land reclamation and (ii) urbanization. The former tends to expand the availability of land to agriculture while the latter constrains it. The net change in land availability is thus the outcome of a rather complex set of decisions. The MISR2 framework does not focus on these issues and allows to shift independently the availability of land according to the best judgement of the analyst.

Each of the production activities has a certain capacity of production defined by its capital stock. This capacity increases via investment which expands the capital stock. At the beginning of each period each activity inherits a new capital stock which is equal to the one it had the preceding period, net of depreciaton, plus any new investment which the activity made. Investment is driving the capital stock. The determination of investment differs between the public and private sectors. As was mentioned, the value of public investment is a decision variable while the private sector invests whatever funds are available to it. The allocation of investment between various public sector activities is again a planning decision, while the allocation of investment in the private sector responds to relative rates of returns.

The MISR2 framework assumes the existence of a rural and an urban labor market. On the former labor supply is assumed given in each period and the rural wage adjusts within the period to clear the market. The labor supply in rural areas shifts between periods because of two factors: (i) the natural rate of growth of the rural labor force and (ii) migration to urban areas. The first factor expands the labor supply and is given exogenously. Migration slows down the growth in the rural labor supply. The amount of migration is determined according to a modified version of the Harris-Todaro explanation (see Dervis, de Melo, Robinson [1982]): the higher the expected urban wage relative to the rural wage, the more important the migration flow will be. Migration to urban areas adds to the natural expansion of the urban labor force to form urban labor supply. Within the period, this labor supply exceeds labor demand at a fixed nominal wage. There is unemployment. The fixed nominal urban wage shifts between periods according to a Philipps curve explanation augmented with inflationary expectations. Thus when the urban unemployment goes up, the increase in the nominal wage slows down. However, the full increase in the urban consumer price index of last period is passed on the wage of the next period. Thus urban real wages are cut by "inflation" within each period but catch up with a lag. They are fully indexed to the "inflation" of previous periods.

Apart from the expansion of land, labor and capital, production activities can see their resource endowments increase through increases in productivity. Hicks-neutral technical change is assumed for all production activities. The rates of technical change are provided exogenously. A higher rate of technical change allows, for a given level of production, a lower marginal cost and hence an increase in the profitability of the activity.

The level of oil production is not strictly speaking a "resource endowment," it is however exogenous. Changes in the output level of oil contribute to the shifts in the general equilibrium of the economy between periods. The growth in oil production has to be assumed according to the best available information.

2. Policy Measures

Decision makers in the MISR2 framework react to a certain number of policy signals. Their reactions, jointly with the constraints imposing the consistency of the outcome of all their decisions, result in an economy-wide equilibrium. Shifts in the policy signals induce agents' reactions and lead to a new equilibrium. Thus policy signals are another set of variables which make the economy move along a sequential equilibrium path.

The policy instruments captured in the MISR2 reference model are the following:

(i) pricing of the production of the public sector activites;

(ii) prices at which the government (GASC) buys goods from domestic agriculture;[1]

(iii) selling prices of the subsidized items;

(iv) levels of rationing of the subsidized commodities;

(v) the exchange rates governing transactions on the central bank and commercial banks foreign exchange pools;

(vi) the wage and employment policies in the government and state enterprises;

(vii) tax rates for both direct and indirect taxation;

1/ GASG is the Government Authority for the supply of commodities.

(viii) government current expenditures on goods and services,

(ix) government investment;

(x) public enterprises overall investment and its allocation,

(xi) net borrowing from abroad through the central and commercial banks pools;

(xii) the allocation of foreign exchange over commodity-imports in the "discretionary" foreign exchange budget."

3. The Rest of the World

A last set of variables which bear on the equilibrium reached in each period and/or its shifts between periods are the rest of the world variables. World prices of both imports and exports and the state of the world economy affect the equilibrium of the Egyptian economy. Outstanding among these variables are: (i) the trends in oil prices; (ii) shifts in the prices of imported subsidized commodities; (iii) the volume of traffic through the Suez Canal; (iv) inflow of tourists and (v) the amounts remitted by Egyptians working abroad. With views on the likely trends in this "world environment," the MISR2 framework allows to derive the macroeconomic and resource allocation implications.

CHAPTER 2: The Formulation of the Inter-Period Module of MISR2

Appendix 1 described the general equilibrium which is obtained within one period. This equilibrium is conditional on resource endowments, policy parameters, structural parameters and the world environment. These variables as discussed in chapter 1 of this appendix define conditions considered fixed within one period but which may vary between periods. Their inter-period variations drive the MISR2 models dynamically. Some of those variations are assumed to follow exogeneously determined trends, others may be the outcome of behavioral decisions and depend on results obtained in previous periods.

In the following we present the specifications governing the non-exogenous inter-period variations. They pertain to investment and capital stocks accumulation on the one hand and to rural-urban migration and urban wage determination on the other hand.

1. Fixed Capital Formation and Capital Stock Accumulation

As mentioned in appendix one and formulated in the previous chapter the capital formation of public sector activities and the government are fixed in nominal terms in each period. Private sector activities are thought to be financially constrained and to invest an amount equal to the value of finance they are able to obtain. The question which arises is that of the allocation of the total funds available to the private sector over the various activities.

One can think of a financial sector which allocates a total amount of resources across various activities. Let $(TFI)_t$ be the total financial resources available to the private sector, then

$$P_{it} I_{it} = \alpha_{it} (TFI)_t, \quad i = 1, 2, \ldots, n \qquad (1)$$

where I_{it} is investment in volume, P_{it} the price of investment, α_{it} is the share of financial resources going to sector i ($\sum_i \alpha_{it} = 1$), and n the total number of private activities. Formally, the question is how are the α_{it}, i = 1, 2, ..., n determined and how do they change over time. Dervis, De Melo and Robinson (1982) suggest the following specification [1]:

$$\alpha_{it} = SP_{i(t-1)} + \mu \, SP_{i(t-1)} \left(\frac{R_{i(t-1)} - AR_{t-1}}{AR_{t-1}} \right), \quad i = 1, 2, \ldots, n \qquad (2)$$

where SP_{it} is a sectoral share in aggregate profits, µ is a parameter indicating intersectoral "mobility" of financial resources, R_{it} is a sectoral profit rate ad AR_t is the average profit rate. Specification (2) indicates that if there is no "mobility" of funds across sectors µ = 0, any difference between the sectoral profit rate and the average profit rate will not affect the share of funds allocated to the sector. Furthermore this share will be equal to the last period share of the sector in aggregate profits. With "mobility" of funds, sectors with profit rates higher than average will pull resources over and above their respective profit shares whereas the sectors with lower than average profit rates would give up resources. Thus with µ > 0, relations (2) allow for allocation shares α_{it} to respond to relative profitability. The variability of the α_{it} in (2) is around sectoral shares in aggregate profits: $SP_{i(t-1)}$. The underlying idea here is that if there were no "mobility" of funds (µ = 0) and if all profits were saved, then the investment undertaken by each activity would be self-financed. Thus (2) is broadly based on the idea that savings come essentially from profits and

---

[1] $0 < SP_{it} < 1$ and $\Sigma SP_{it} = 1$ hence $\Sigma \alpha_{it} = 1$ for all µ's.

that firms get more or less than their available resources according to their relative profitability and to the "mobility" of funds. However, in MISR2 relation 2 would be applied to shares allocating funds available to private activities between the latter. Hence the link between the funds which each activity would receive, its profits, and eventual savings is more remote than what is implicit in (2). Furthermore, the mechanism assumed in MISR2 is that the financial sector allocates across private activities a given amount of funds. It seems difficult to assume that such a financial sector would be able to have a good idea of aggregate private profits and hence of an average profit rate in the whole private economy, weighted by sectoral profit shares. A reasonable approach would be to assume: (i) that the financial sector knows the allocations of funds it has made in past periods, (ii) that it is provided with individual profit rates for each private activity, and (iii) that based on previous allocations of funds, it uses available profit rates to fix an average profit rate for its activities in the private sector. These assumptions would lead to a respecification of (2):

$$\alpha_{it} = \beta_{i(t-1)} + \mu \beta_{i(t-1)} \left(\frac{R_{i(t-1)} - \overline{R}_{(t-1)}}{\overline{R}_{(t-1)}} \right), \quad i = 1, 2, \ldots, n, \quad (3)$$

where $\beta_{i(t-1)}$ are the shares used in last period to allocate the funds, $R_{i(t-1)}$ are sectoral profit rates, μ is still a "mobility" of funds parameter and

$$\overline{R}_{(t-1)} = \sum_{i=1}^{n} \beta_{i(t-1)} R_{i(t-1)}. \quad (4)$$

Given (4), (2) obviously satifies the activity constraint $\sum_i \alpha_{it} = 1$. Relations (3) tell the following story: the financial sector has some inertia in its allocation of funds, however at each period it reviews the performance of private activities and according to the extent of the "mobility" of funds it reorients its financing between various private activities. To complete the specification of (3), one needs to define more precisely the sectoral profit rates R_i. Let $P_{i(t-1)} K_{i(t-1)}$ be the value of the capital stock, in sector i, at the end of period t-1, let δ_i be the rate of depreciation in the sector i and OS_{it} be the gross operating surplus in sector i, in period t, the profit rate R_{it} is defined as [1]:

---

[1] This formulation of R_{it} ignores capital gains.

$$R_{it} = \frac{OS_{it} - P_{i(t-1)} \delta_i K_{i(t-1)}}{P_{i(t-1)} K_{i(t-1)}}, \quad i = 1, 2, \ldots, n \tag{5}$$

Once investment decisions are made, the capital endowements for the next period are derived. This is done in the same way for all production activities whether in the private or public sectors. Let I_{jt} by the amount of capital formation going into sector j, then:

$$K_{j(t+1)} = I_{jt} + (1 - \delta_j) K_{jt}, \tag{6}$$

where K_j is the capital stock of sector j and δ_j is a rate of depreciation which is sector specific. The index t is for time.

2. Rural-Urban Migration and Urban Wage Determination

Two labor markets are captured in MISR2: rural and urban labor markets. Rural households supply labor services to the former and urban households to the latter. Migration is assumed to take place from rural to urban areas and is described by the following equation:

$$MIG_t = \gamma \{ \frac{W_u^e}{W_R} - 1 \} L_{Rt}, \tag{7}$$

where W_u^e is the expected urban wage [1] and W_R is the rural wage. The higher the parameter $\gamma > 0$ and the stronger is the response of migration to the wage differential. In (7) L_{Rt} is the rural labor supply. The formulation (7) is a variation on the Harris-Todaro version since it does require the equalization of the expected urban wage with the rural wage. Rural labor supply will be given by:

$$L_{Rt} = (1 + g_R) L_{R(t-1)} - MIG_{t-1}, \tag{8}$$

---

[1] $W_u^e = W_u(1-p)$ where W_u is the actual urban wage and p the urban unemployment rate. Specification (7) is also used in the Domestic Resource Mobilization model of Egypt (1980).

References

Agarwala, R. (1983), <u>Price, Distortions and Growth in Developing Countries,</u> World Bank Staff Working Paper No. 575, Washington, D.C.

Ahmed, S. (1984), <u>Industrialization in Egypt: Performance and Policies,</u> mimeo, World Bank, Washington, D.C.

Amranand, P. and Grais, W. (1984), <u>Macroeconomic and Distributional Implications of Sectoral Policy Interventions: An Application to Thailand,</u> World Bank Staff Working Paper No. 627, Washington, D.C.

Blitzer, C.R. Clark, P.B. and Taylor, L. (1975), <u>Economy-Wide Models and Development Planning,</u> Oxford University Press, London.

Carlevaro, F. (1975), <u>Sur la Comparaison et la Generalisation de Certains Systemes de Functions de Consommnation Semi-Agregees,</u> Peter Lang, Bern.

Chewa-Krengkai, A. and Lamsam B. (1982), <u>A Social Accounting Matrix for Thailand 1975,</u> Joint Publication by NESDB and IBRD, Bangkok and Washington, D.C.

Deaton, A. and Muellbauer, J. (1980), <u>Economics of Consumer Behavior,</u> Cambridge University.

Dervis, K., de Melo J., and Robinson, S. (1982), <u>General Equilibrium Models for Development Policy,</u> Cambridge University Press, Cambridge.

Dewatripont, M., Michel, G. (1983), <u>On Closure Rules and Micro-Foundations in Applied General Equilibrium Models,</u> World Bank, mimeo.

Drud, A., Grais, W. and Pyatt, G. (1983), <u>The TV-Approach: A Systematic Method of Defining Economy-Wide Marco Models Based in Social Accounting Matrices,</u> mimeo, World Bank, Washington, D.C.

Drud, A., Grais, W. and Vujovic, D. (1982), <u>Thailand: An Analysis of Structural and Non-Structural Adjustments,</u> World Bank Staff Working Paper No. 513, Washington, D.C.

Ginsburgh and Robison, D. (1983), <u>Equilibrium and Prices in Multisector Models,</u> mimeo, World Bank, Washington, D.C.

King, B. (1981), <u>What is a SAM? A Layman's Guide to Social Accounting Matrices,</u> World Bank Staff Working Paper No. 463, Washington, D.C.

Lysy, F. (1982), <u>The Character of General Equilibrium Models under Alternative Closure Rules,</u> World Bank, mimeo.

Neary, J.P. and Robert, K.W.G. (1980), "The Theory of Household Behavior Under Rationing", <u>European Economics Review,</u> Vol 13.

rate of unemployment p tends towards the frictional rate $\bar{p}$, the u
induced rate of increase in wages will tend to α_0:

$$\lim_{p \to \bar{p}} z = \alpha_0;$$

similarly when the rate of unemployment increases substantially,
gradually decrease to $\alpha_0 - \alpha_1$. The parameter β will indicate [1]
which z varies between α_0 and $\alpha_0 - \alpha_1$. Thus when the rate of une
increases the pressure for the increase of the urban wage will di
the contrary when the urban labor market is tight a maximum rate
increase (α_0) is approached. The expression between brackets in
additional term (ν) representing "inflationary" expectation. A s
specification reflecting the full adjustment of the wage to the ":
last period is adopted for ν :

$$\nu_t = \frac{PC_t}{PC_{t-1}} - 1 ,$$

where PC_t is the urban CPI and t is a time index. [2]

---

[1] Naturally Max {p} = 1. Relation (12) is a downward sloping lc
an inflexion point at $\beta/2$.

[2] ν_t indicates the inflationary expectation held at the end of p
period t + 1.

where g_R is the actual rate of growth of the rural labor force.[1] Similarly, the urban labor supply will be:

$$L_{ut} = (1 + g_u) L_{u(t-1)} + MIG_{t-1}, \qquad (9)$$

where g_u is the natural rate of growth of the urban labor force. Given labor demand on the urban areas, using (9) one can derive unemployment and the rate of unemployment:

$$p = \frac{L_{ut} - LD_{ut}}{L_{ut}}, \qquad (10)$$

where LD_{ut} is urban labor demand and p is the urban rate of unemployment.

In the within-period reference model of the MISR2 class it is assumed that urban labor supply is perfectly elastic at the going wage. This implies the existence of unemployment as described by (10) and assumes a fixed urban wage in each period. That wage is allowed to change between periods according to the level of the rate of unemployment and to inflationary expectations consistent with a Philips curve explanation augmented with inflationary expectations:

$$W_t = W_{t-1} (1 + z + \nu) \qquad (11)$$

where z is a function of the rate of unemployment and ν is the expected rate of increase of the urban Consumer Price Index (CPI). The specification chosen for z is:

$$z = \alpha_0 - \alpha_1 \exp[-\beta/(p - \bar{p})], \qquad (12)$$

where $0 < \alpha_0 < 1$, $0 < \alpha_1 < 1$ and $\beta > 0$ are parameters and $\bar{p}$ the frictional rate of unemployment corresponding to a tight urban labor market. When the

---

[1] It is possible to adjust relation (8) and (9) to take into account international labor migration. Here it is assumed that net migration in the eighties will not have a major impact.

Pleskovic, B. and Croswell, M. (1981), Social Accounting Matrices for Egypt: Outlines and Suggestions for Disaggregation of Individual Accounts, mimeo, World Bank, Washington, D.C.

Pyatt, G., and Thorbecke, E. (1976), Planning Techniques for a Better Future, International Labor Organization, Geneva.

Pyatt, G., Roe, A., et al (1977), Social Accounting for Development Planning: With Special Reference to Sri Lanka, Cambridge University Press, Cambridge, U.K.

Taylor, L. (1979), Macro Models for Developing Countries, MacGraw Hill, New York.

United Nations (1968), A System of National Accounts, Series F, No. 2, Rev. 3, New York.

World Bank (1983), Arab Republic of Egypt - Current Economic Situation and Growth Prospects, Report No. 4498-EGT, Washington, D.C.

World Bank (1983), Arab Republic of Egypt - Issues of Trade Strategy and Investment Planning, Report No. 4136-EGT, Washington, D.C.

World Bank (1980), Arab Republic of Egypt - Domestic Resource Mobilization and Growth Prospects for the 1980s, Report No. 3123-EGT, Washington, D.C.

World Bank Publications of Related Interest

Adjustment Experience and Growth Prospects of the Semi-Industrial Countries
Frederick Jaspersen

Staff Working Paper No. 477. 1981. 132 pages (including 3 appendixes).
Stock No. WP 0477. $5.

Adjustment in Low-Income Africa
Robert Liebenthal

Staff Working Paper No. 486. 1981. 62 pages (including bibliography).
Stock No. WP 0486. $3.

Aggregate Demand and Macroeconomic Imbalances in Thailand: Simulations with the SIAM 1 Model
Wafik Grais

Staff Working Paper No. 448. 1981. 132 pages (including 3 appendixes).
Stock No. WP 0448. $5.

NEW

Alternative Mechanisms for Financing Social Security
Parthasarathi Shome and Lyn Squire

Reviews, clarifies, and evaluates theoretical literature about the effect of social security on capital accumulation and labor supply. Analyzes empirical studies using U.S. data, the impact of pay-as-you-go financed and fully funded social security schemes, and characteristics of optimal social security systems. This study provides a starting point for everyone interested in the relevance of existing theories for financing social security in developing countries.

Staff Working Paper No. 625. 1983. 62 pages.
ISBN 0-8213-0292-2. Stock No. WP 0625. $3.

An Analysis of Developing Country Adjustment Experiences in the 1970s: Low-Income Asia
Christine Wallich

Staff Working Paper No. 487. 1981. 43 pages (including references).
Stock No. WP 0487. $3.

Aspects of Development Bank Management
William Diamond and V. S. Raghavan

Deals exclusively with the management of development banks. The book is divided into eight sections, each dealing with one aspect of management of its problems, and of the various ways of dealing with them.

EDI Series in Economic Development. The Johns Hopkins University Press, 1982. 2nd printing, 1983. 311 pages.
LC 81-48174. ISBN 0-8018-2571-7, Stock No. JH 2571, $29.95 hardcover; ISBN 0-8018-2572-5, Stock No. JH 2572, $12.95 paperback.

Capital Accumulation in Eastern and Southern Africa: A Decade of Setbacks
Ravi Gulhati and Gautam Datta

Analyzes the magnitude of the setback in capital accumulation in eastern and southern Africa. This phenomenon is examined in twenty-eight statistical tables. The authors sample sixteen countries and rely on expert observations to explore the proximate causes of the setbacks.

World Bank Staff Working Paper No. 562. 1983. 74 pages.
ISBN 0-8213-0169-1. Stock No. WP 0562. $3.

Capital Market Imperfections and Economic Development
Vinayak V. Bhatt and Alan R. Roe

Staff Working Paper No. 338. 1979. 87 pages (including footnotes).
Stock No. WP 0338. $3.

The Changing Nature of Export Finance and Its Implications for Developing Countries
Albert C. Cizauskas

Staff Working Paper No. 409. 1980. 43 pages (including 3 annexes).
Stock No. WP 0409. $3.

NEW

Compounding and Discounting Tables for Project Analysis (with a Guide to Their Applications)
Second Edition, Revised and Expanded
J. Price Gittinger

Project planners and analysts will find this book a convenient and time-saving reference for the preparation and analysis of development projects. Six-decimal tables for 1 percent through 50 percent show the compounding factor for 1 and for 1 per annum, the sinking fund factor, the discount factor, the present worth of an annuity factor, and the capital recovery factor. The first edition of this book underwent seven printings in ten years and was translated into Arabic, Chinese, French, and Spanish. This new edition—with narrow-interval compounding tables added for higher interest rates, updated project examples, a guide to using simple electronic calculators to perform the computations discussed, and an annotated bibliography increases the proven usefulness of its predecessor, both in the classroom and at the project site.

May 1984. About 208 pages.
ISBN 0-8018-2409-5. Stock No. BK 2409. $10.95.

Translations of this new edition will be available in 1985. Still available are the following translations of the first edition:

French: *Tables d'interets composés et d'actualisation. Economica, 4th printing, 1979.*
ISBN 2-7178-0205-3, Stock No. IB 0542, $6.

Spanish: *Tablas de interes compuesto y de descuento para evaluación de proyectos. Editorial Tecnos, 1973; 4th printing, 1980.*
ISBN 84-309-0716-5, Stock No. IB 0526. $6.

A Conceptual Approach to the Analysis of External Debt of the Developing Countries
Robert Z. Aliber

Staff Working Paper No. 421. 1980. 25 pages (including appendix, references).
Stock No. WP 0421. $3.

NEW

Development Finance Companies, State and Privately Owned: A Review
David L. Gordon

An informative guide to the function and design of development finance companies as they are set up in developing countries. Case histories highlight the differences among these companies—their institutional structure, management style, financial performance, and other features. Looks at the problems of resource mobilization and strategies to overcome them.

Staff Working Paper No. 578. 1983. 84 pages.

ISBN 0-8213-0226-4. Stock No. WP 0578. $3.

Development Prospects of Capital Surplus Oil-Exporting Countries: Iraq, Kuwait, Libya, Qatar, Saudi Arabia, UAE
Rudolf Hablützel

Staff Working Paper No. 483. 1981. 53 pages (including statistical tables).

Stock No. WP 0483. $3.

Developments in and Prospects for the External Debt of the Developing Countries: 1970-80 and Beyond
Nicholas C. Hope

Staff Working Paper No. 488. 1981. 70 pages (including 2 annexes, references).

Stock No. WP 0488. $3.

NEW

Domestic Resource Mobilization in Pakistan: Selected Issues
Nizar Jetha, Shamshad Akhtar, and M. Govinda Rao

Fouses on the relationship between taxation and the three main components of savings. Emphasizes tax reform with a view to raising additional revenues and encouraging household and business savings. Proposals for tax reform take account of equity considerations and the need to keep tax-induced distortions in the allocation of resources to a minimum. Highlights appropriate policies on current expenditures, subsidies, user charges, public enterprise pricing, self-financing of investment by public enterprises. Includes three annexes that examine direct taxes, indirect taxes, and tax changes in Pakistan's 1983/84 budget.

Staff Working Paper No. 632. 1984. 144 pages.

Stock No. WP 0632. $5.

NEW

Economic Liberalization and Stabilization Policies in Argentina, Chile, and Uruguay: Applications of the Monetary Approach to the Balance of Payments
Edited by Nicolas Ardito Barletta, Mario I. Blejer, and Luis Landau

Twenty-eight leading international economists and regional specialists review the salient characteristics of the monetary approach to the balance of payments, examine the variations in its application, and evaluate its successes and failures. Emphasizes the empirical evidence and dynamic aspects and costs. Provides an important examination of economic policies and their effects in a region that looms large in current deliberations about international indebtedness and finance.

June 1984. About 240 pages.

ISBN 0-8213-0305-8. $17.50 paperback.

Energy Prices, Substitution, and Optimal Borrowing in the Short Run: An Analysis of Adjustment in Oil-Importing Developing Countries
Ricardo Martin and Marcelo Selowsky

Staff Working Paper No. 466. 1981. 77 pages (including footnotes, references).

Stock No. WP 0466. $3.

Exchange Rate Adjustment under Generalized Currency Floating: Comparative Analysis among Developing Countries
Romeo M. Bautista

Staff Working Paper No. 436. 1980. 99 pages (including appendix).

Stock No. WP 0436. $3.

A General Equilibrium Analysis of Foreign Exchange Shortages in a Developing Economy
Kemal Dervis, Jaime de Melo, and Sherman Robinson

Staff Working Paper No. 443. 1981. 32 pages (including references).

Stock No. WP 0443. $3.

Prices subject to change without notice and may vary by country.

Growth and Structural Adjustment in East Asia
Parvez Hasan

Staff Working Paper No. 529. 1982. 42 pages.

ISBN 0-8213-0102-0. Stock No. WP 0529. $3.

Interest Rate Management in Developing Countries: Theory and Simulation Results for South Korea
Sweder van Wijnbergen

Examines the claim that higher time deposit rates raise output and lower inflation in the short run, and increase growth through their favorable impact on savings rates. It concludes that this theory depends heavily on the assumption that portfolio shifts into time deposits come out of unproductive assets, providing less intermediation than the banking system. Impact of changes in time deposit rates on inflation, capital, capital accumulation and medium term growth are discussed, and empirical relevance is demonstrated through simulation runs with a macroeconomic model of South Korea.

World Bank Staff Working Paper No. 593. 1983. 52 pages.

ISBN 0-8213-0188-8. Stock No. WP 0593. $3.

International Adjustment in the 1980s
Vijay Joshi

Staff Working Paper No. 485. 1982. 57 pages.

ISBN 0-8213-0062-8. Stock No. 0485. $3.

NEW

Links between Taxes and Economic Growth: Some Empirical Evidence
Keith Marsden

Reviews the experience with growth and taxation in twenty developing and developed countries, spanning a wide spectrum of incomes. Do countries with lower taxes experience more rapid expansion of investment, productivity, employment, and government services? This provocative paper sheds new light on this and other key questions especially relevant to development economists. It also examines the mechanisms by which fiscal policies may affect growth rates.

Staff Working Paper No. 605. 1983. 48 pages.

ISBN 0-8213-0215-9. Stock No. WP 0605. $3.

NEW

Municipal Accounting for Developing Countries
David C. Jones

This manual is based on British practices and terminology of municipal accounting, modified to suit the needs of other countries, especially those lacking a core of appropriately trained accountants. Provides the basic principles of municipal accounting for those with little or no bookkeeping experience and proceeds through successive levels of difficulty to some of the most advanced concepts currently in use, including the pooling of loans. An important feature is the multitude of practical applications and examples of forms and records.

A joint publication of the Chartered Institute of Public Finance and Accountancy and the World Bank.

June 1984. About 900 pages.

ISBN 0-8213-0350-3. Stock No. BK 0350. $30.

The Nature of Credit Markets in Developing Countries: A Framework for Policy Analysis
Arvind Virmani

Staff Working Paper No. 524. 1982. 204 pages.

ISBN 0-8213-0019-9. Stock No. WP 0524. $5.

The Newly Industrializing Developing Countries after the Oil Crisis
Bela Balassa

Staff Working Paper No. 437. 1980. 57 pages (including appendix).

Stock No. WP 0437. $3.

Notes on the Analysis of Capital Flows to Developing Nations and the "Recycling" Problem
Ralph C. Bryant

Staff Working Paper No. 476. 1981. 67 pages.

Stock No. WP 0476. $3.

Notes on the Mechanics of Growth and Debt
Benjamin B. King

A practical model to explore the way in which capital inflow from abroad affects economic growth.

The Johns Hopkins University Press, 1968. 69 pages (including 4 annexes).

LC 68-8701. ISBN 0-8018-0338-1, Stock No. JH 0338. $5 paperback.

The Policy Experience of Twelve Less Developed Countries, 1973-1978
Bela Balassa

Staff Working Paper No. 449. 1981. 36 pages (including appendix).

Stock No. WP 0449. $3.

The Political Structure of the New Protectionism
Douglas R. Nelson

Staff Working Paper No. 471. 1981. 57 pages (including references).

Stock No. WP 0471. $3.

NEW

Price Distortions and Growth in Developing Countries
Ramgopal Agarwala

Sixteen informative tables trace the distortion in prices of foreign exchange and other factors affecting the growth of developing countries. Based on statistics from thirty-one developing countries.

Staff Working Paper No. 575. 1983. 78 pages.

ISBN 0-8213-0242-6. Stock No. WP 0575. $3.

Pricing Policy for Development Management
Gerald M. Meier

Presupposing no formal training in economics, it explains the essential elements of a price system, the functions of prices, the various policies that a government might pursue in cases of market failure, and the principles of public pricing of goods and services provided by government enterprises. It also provides the would-be practitioner with an appreciation of the underlying logical structure of cost-benefit project appraisal. To give substance to the applied and policy dimensions, many of the readings are drawn from the experience of development practitioners and relate to such important sectors as agriculture, industry, power, urban services, foreign trade, and employment. The principles outlined are therefore relevant to a host of development problems.

The Johns Hopkins University Press. 1983. 272 pages (including bibliography and index).

LC 82-7716. ISBN 0-8018-2803-1, Stock No. JH 2803, $35 hardcover; ISBN 0-8018-2804-X, Stock No. JH 2804, $12.95 paperback.

Private Bank Lending to Developing Countries
Richard O'Brien

Staff Working Paper No. 482. 1981. 60 pages (including appendix, bibliography).

Stock No. WP 0482. $3.

Private Capital Flows to Developing Countries and Their Determinations: Historical Perspective, Recent Experience, and Future Prospects
Alex Fleming

Staff Working Paper No. 484. 1981. 41 pages.

Stock No. WP 0484. $3.

Private Direct Foreign Investment in Developing Countries
K. Billerbeck and Y. Yasugi

Staff Working Paper No. 348. 1979, 101 pages (including 2 annexes).

Stock No. WP 0348. $5.

NEW

Savings Mobilization through Social Security: The Case of Chile, 1916–1977
Christine Wallich

Describes the savings mobilization potential in Chile and in five Asian programs. Some sort of social security program functions in almost all developing countries. Programs are often costly, whether measured in relation to GNP, government expenditure, government revenue, or the wage bill. This paper compares the successful systems.

Staff Working Paper No. 553. 1983. 109 pages.

ISBN 0-8213-0123-3. Stock No. WP 0553. $5.

Short-Run Macro-Economic Adjustment Policies in South Korea: A Quantitative Analysis
Sweder van Wijnbergen

Staff Working Paper No. 510. 1981. 182 pages (including 3 appendixes).

ISBN 0-8213-0000-8. Stock No. WP 0510. $5.

Prices subject to change without notice and may vary by country.

State Finances in India

A three-volume set of papers that explores a range of issues relating to the nature of intergovernmental fiscal relations in India.

Vol. I: Revenue Sharing in India
Christine Wallich

Vol. II: India—Studies in State Finances
Christine Wallich

Vol. III: The Measurement of Tax Effort of State Governments, 1973-1976
Raja J. Chelliah and Narain Sinha

Staff Working Paper No. 523. 1982. vol. I, 85 pages, vol. II, 186 pages, vol. III, 85 pages.

ISBN 0-8213-0013-X. vol. I, Stock No. WP 1523, $3, vol. II, Stock No. WP 2523, $5, vol. III, Stock no. WP 3523, $3.

Structural Adjustment Policies in Developing Economies
Bela Balassa

Staff Working Paper No. 464. 1981. 36 pages.

Stock No. WP 0464. $3.

Structural Aspects of Turkish Inflation: 1950-1979
M. Ataman Aksov

Staff Working Paper No. 540. 1982. 118 pages.

ISBN 0-8213-0098-9. Stock No. WP 0540. $5.

Thailand: An Analysis of Structural and Non-Structural Adjustments
Arne Drud, Wafik Grais, and Dusan Vujovic

Staff Working Paper No. 513. 1982. 93 pages (including appendix).

ISBN 0-8213-0023-7. Stock No. WP 0513. $3.

Trends in Rural Savings and Private Capital Formation in India
Raj Krishna and G.S. Raychaudhuri

World Bank Staff Working Paper No. 382. 1980. 43 pages (including 2 tables, 3 appendixes, references).

Stock No. WP 0382. $3.

"The primary source for medium- and long-term external debt of many developing countries."

Suhas Ketkar, Asia-Pacific Economist and Vice President, Marine Midland Bank, N.A.

"Often the only reliable source of information for countries for which data is hard to come by . . . Used quantitatively for macroeconomic detail as well as qualitatively in reports discussing the debt picture. I find the projected servicing payments a strong feature."

Jonathan Kayes, International Economist, Republic National Bank of New York

World Debt Tables, 1983-84 Edition

The World Bank's invaluable reference guide to the external debt of developing countries. Essential planning tool for economists, bankers, country risk analysts, financial consultants and all those interested in the global system of trade and payments. Provides data on the external debt of 103 developing countries augmented by information, where available, on major economic aggregates plus indicators used to analyze debt and creditworthiness. Shows statistical tables by country, including figures for external public debt outstanding, commitments, disbursements, service payments, and net borrowings. Reports on private nonguaranteed debt of 19 countries. Gives aggregate position of 13 major borrowers—countries with disbursed and outstanding medium- and long-term total debt in excess of $13.5 billion at the end of 1982. Includes periodic supplements as fresh data are received.

1984. 328 pages.

Stock No. BK 0315 $75 (annual subscription)

Also available for the first time . . .
Summary Report

Debt and the Developing World: Current Trends and Prospects

Includes an overview and summary tables from the 1983-84 edition.

1984. 64 pages.

Stock No. BK 0319, $6.50.

Companion computerized data base

Includes all debt information given in the unabridged volume, and, where available, offers continuous historical series for 1970-82 and projected debt-service payments for 1983-92. Write for sample purchase agreement.

(9-track, phase-encoded, recording density 1600 bpi)

Stock No. IB 0500, $5,000 (service bureaus for reselling to their clients); Stock No. IB 0667, $2,000 (banks and commercial corporations); Stock No. IB 0666, $500 (universities and libraries).